空间美学

SPACE AESTHETICS

深圳视界文化传播有限公司 编
Shenzhen Design Vision Cultural Dissemination Co., Ltd

中国林业出版社
China Forestry Publishing House

DESIGN METHODS ABOUT STRATEGY FIRST
策略为先的设计方法

My friends often ask where my design inspirations come from, and I always tell them tongue-in-cheek that “inspirations fall down from above”. Facing with so many designs, inspirations may be late. But as professional designers, we need to find a way to solve the problem in the absence of inspirations.

Sometimes I’m worried because I cannot find inspirations, but I’ll warn myself it is because the thoughts are not thorough enough or the researches are not meticulous enough that I cannot find inspirations. So I think if designers can improve the solutions to the state of strategies or tactics, there must be more creative thoughts.

I often talk with my colleagues that don't rush to design when getting a project, assess the situation as Zhuge Liang did and take the sky, earth and people into consideration.

The sky is the climate and season. There are great climatic differences between the south and north, so it has a great influence on the space planning of the architecture. The climate of Southeast Asian area is hot, so many resort hotels are open. While in the north, the structures of architecture are closed to better defense the cold winter. At the same time, we need take the seasons into consideration. A show flat built in different seasons may have different concerns on strategy.

The earth is about the region, area, environment and space. From the south, north, east and west, different regions can bring different climate. In the same city, the economic development of different areas is different. People who live in the west are the noble, the south are the rich and the north are the poor. These elements can influence judgments of the designers. What’s more, the style, appearance and space structure of a city can have a great influence on the interior design.

People are about humanity, habit, style and aesthetics, cost estimate. Different people have different aesthetics, and people from different area have aesthetic consciousness concentrated by their areas. If it is a private residence, it may have closer relations with human characteristics. Different people have different concepts of home.

“Strategy first and temperament above” is our design belief. Only concerning strategy first can make more outstanding works.

序言 PREFACE

朋友常问我的设计灵感从何而来，我总是半开玩笑地告诉他们："灵感是从天上掉下来的。"面对很多设计时，灵感可能总会迟到，但作为一个职业设计师，我们必须找到一种方法，在没有灵感的时候用它解决问题。

有时我会为找不到灵感而发愁，但我会告诫自己：找不到设计灵感一定是因为思考不够透彻，或是项目研究还不够细致。所以我想如果设计师可以把解决方案提高到策略甚至战略的高度，那么将会有更多创造性思维出现。

我常和同事们说，拿到一个项目不要急于下笔设计，一定要像诸葛亮一样审时度势，将天、地、人这几个要素作为设计重要的输入条件。

天，就是气候季节。在南方和北方，气候条件差异较大，因而会对建筑的空间格局产生很大的影响。东南亚地区气候炎热，所以很多度假型酒店基本上都是比较开敞的。而在北方，很多建筑的格局比较封闭，则是为了更好地阻挡冬天凛冽的寒风。同时也要将季节纳入到设计考虑当中，一个样板房在不同的季节开放，都会有一些策略上的不同考虑。

地，是关乎地域、区域、环境以及空间因素的考虑。比如南面、北面、东面、西面，地域的不同会带来气候的不同，同一个城市不同区域的经济发展状况参差不齐，比如西面住的是贵人、南面住的是富人、北面住的是穷人，这些因素都会影响设计师的判断。再如一个城市的风格风貌、空间结构也是会对室内设计产生重要的影响。

人，是关于人文、习惯、风格审美、造价预算等和人相关的因素，不同的人会有不同的审美，不同区域的人会有他们区域集体的审美意识。如果做私人住宅，那么跟人的特质会更加密切相关，不同的人必定对家有不同的概念。

"策略为先，气质为上"是我们设计的信条，只有让策略先行，才能设计出更为出众的作品。

Li Yizhong
Owner & Creative Director of Li Yizhong Interior Design and DSSY Residential Interior Design
李益中
李益中空间设计、都市上逸住宅设计设计总监

目录
CONTENTS

FRENCH STYLE 法式风格

SIMPLE EUROPEAN STYLE 简欧风格

NEO-CHINESE STYLE 新中式风格

MODERN STYLE 现代风格

AMERICAN STYLE 美式风格

法式风格

FRENCH STYLE

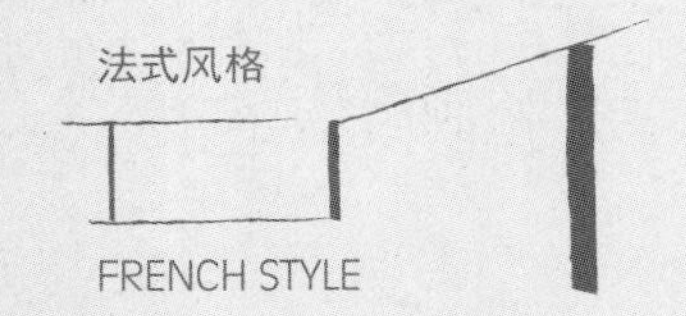

DESIGN CONCEPT 设计理念

Originated from France, the neo-classical style keeps elegant and dignified temperament of classicism and presents a magnificent manner because of opposing rococo furnishings and simplifying traditional style. This project has an area of 700 square meters with three floors on the ground and two floors underground. In addition with the position of products and the value of itself, the design of the villa is defined as French mansion.

Large areas of wood veneer decorative lines and simplified decorative elements with exquisite details create an elegant and high-quality space. Compared with shallow white, the superior gray can easily create an exalted atmosphere. Gray carpet, gray wall, gray plaque and gray cloth are used in the gray and wood color space, while the dynamic orange suddenly makes the space vivid. The designer's control of details is perfect. The exposures of the patterns, such as plum blossom and animals of the carpet in the bedroom can echo with marble texture of the ground and wall. The crystal droplight in the aisle sets off the patterns in the ceiling, which is meticulous.

HOUSE IN GORGEOUS COLORS

锦色春秋

项目名称：万科观承别墅样板房	设计公司：IADC 涞澳设计	设计师：潘及
项目地点：北京	项目面积：600 平方米	
主要材料：大理石、木饰面、布艺等		

新古典风格起源于法国，既保留了古典主义典雅端庄的气韵，又因反对洛可可的矫饰而对传统进行了改良简化，呈现出蔚为大气的风范。针对本案地上三层、地下两层，总量 700 多平方米的体量，以及产品的定位与自身的价值，别墅的设计被顺理成章地定义成了法式大宅。

运用大面积木饰面装饰线条，以及简化了的却有着精致细节的装饰元素，成功塑造了空间典雅的品质之感。“相对于白色的肤浅，高级的灰调更容易营造尊贵的气氛，”灰色的地毯、灰色的墙面、灰色的饰板、灰色的布艺等，同时在灰色与木色之中，一抹动感的橙色则令整个空间一下子生动起来。设计师对于细节的把控更是堪称极致，哪怕是卧室地毯上梅花、动物等图案的露出位置，与地面、墙面石材肌理相呼应，走道水晶吊灯映衬于天花处的花纹，都一丝不苟。

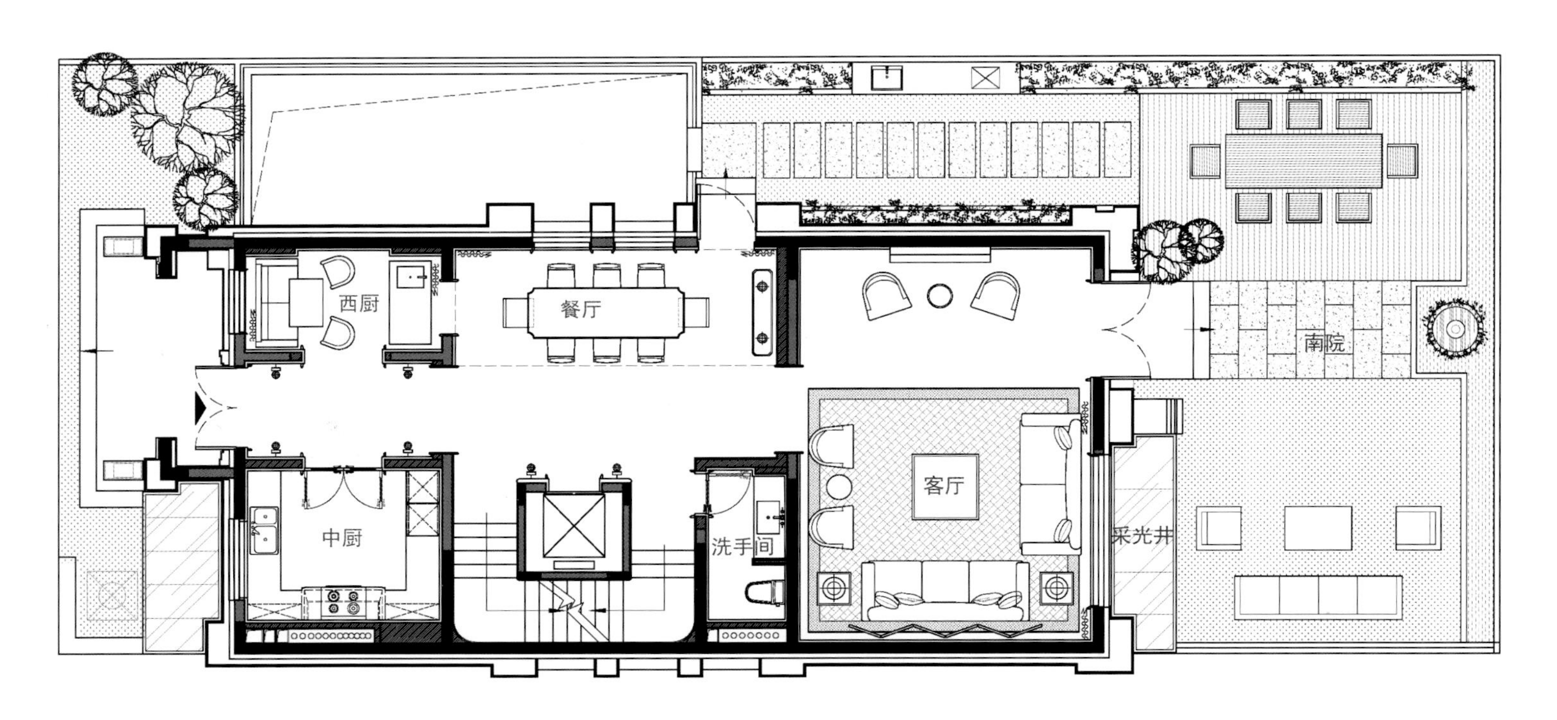

首层的 182.8 平方米空间是家庭的公共区域，一切都围绕着公共核心区的客厅展开，包括由独立的中厨区域、惬意的西厨（下午茶）区域、挑空开放的用餐区域组成的餐厨空间。

首层的空间布局采用对称的格局，意在体现大宅的风范。以客厅视觉端头的英国手工古典壁炉为中心点，结合立面带有东方风格的建筑手绘壁纸，体现设计中欧式古典与东方情怀的融合，将空间从单纯的装饰上升为文化艺术的层次，既没有降低法式情调的主旨，更体现了法国当年崇尚东方元素的潮流。呼应软装饰中的新古典家具，设计师搭配了一些纯粹欧洲古典的家具点缀其中，别具风格的感观跃然而出，不仅有别于同类住宅的风格定位，也体现了业主的艺术文化品味。

客厅背景墙上精心挑选的装饰画，并非简单的摄影作品，而是精心选取的欧洲古典建筑精华，专门央人绘制，与空间陈设相得益彰。华丽的水晶吊灯，不仅增添了空间的装饰感，更提示了各个空间高敞的格局，拉升了空间在视觉上的纵深感。

餐厨空间三段式的格局设计是本案的一个华彩。隔出的三进空间，各司其职，既满足了生活所需，又充分地利用了空间，一举多得。穿过餐厅与客厅的出口，那里正是有着 87 平方米的别墅后花园，花园带有水景及休闲区域，还专门配备了 BBQ 区域，不仅满足了业主日常的家庭活动，更可用于聚会所需。餐厅出口处的天井区域，设计师增加了一个纵向的空间通道，可通过户外的楼梯通道到达地下两层的下沉庭院。

二层的区域为 105.19 平方米，以人物设定的故事为主线，定义了家庭各成员的私人空间，分别为老人房、女孩房及男孩房。整层的布局都为空间的使用者安置了衣帽间及相应的设施，完全满足了日常的生活起居。为凸显对老人的关怀，老人的卧室设定为套房的格局，拥有独立衣帽间及卫生间，令日常的起居更加便捷。女孩房延续了空间灰色配橙色的主调，却增强了装饰的细节，提升了少女所需的唯美度。男孩房更是大胆地融入了许多黑色的细节，以彰显少年青春期的叛逆，书桌区域神来一笔地将少年刻画成一个摄影爱好者，为所有的器材设置了摆放的位置，并合理布局工作区，背墙上更是装饰了一片装饰板架，用于展示他自己及欣赏的摄影作品。所有的软装搭配均符合了使用者特定的人物设定，突出人物的年龄、性格和爱好，加强了空间的装饰个性。

三层空间共计 99.14 平方米为主卧及配套的独立私人空间。24 平方米的卧室、12 平方米的卫生间以及 23 平方米开敞通透的衣帽间中，依然运用了统一的装饰手法，突出法式新古典风格的装饰元素以及固有的庄重与品质。

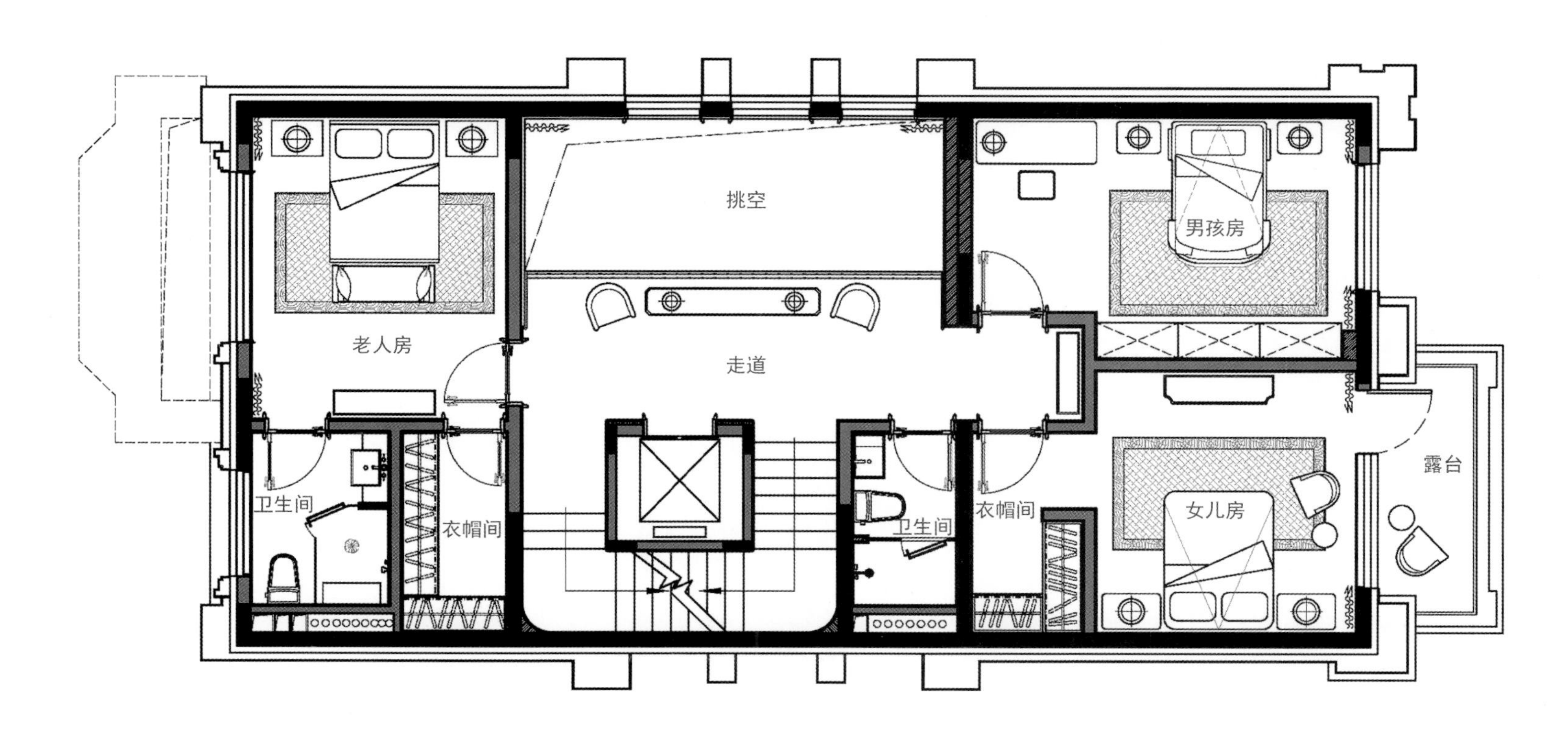

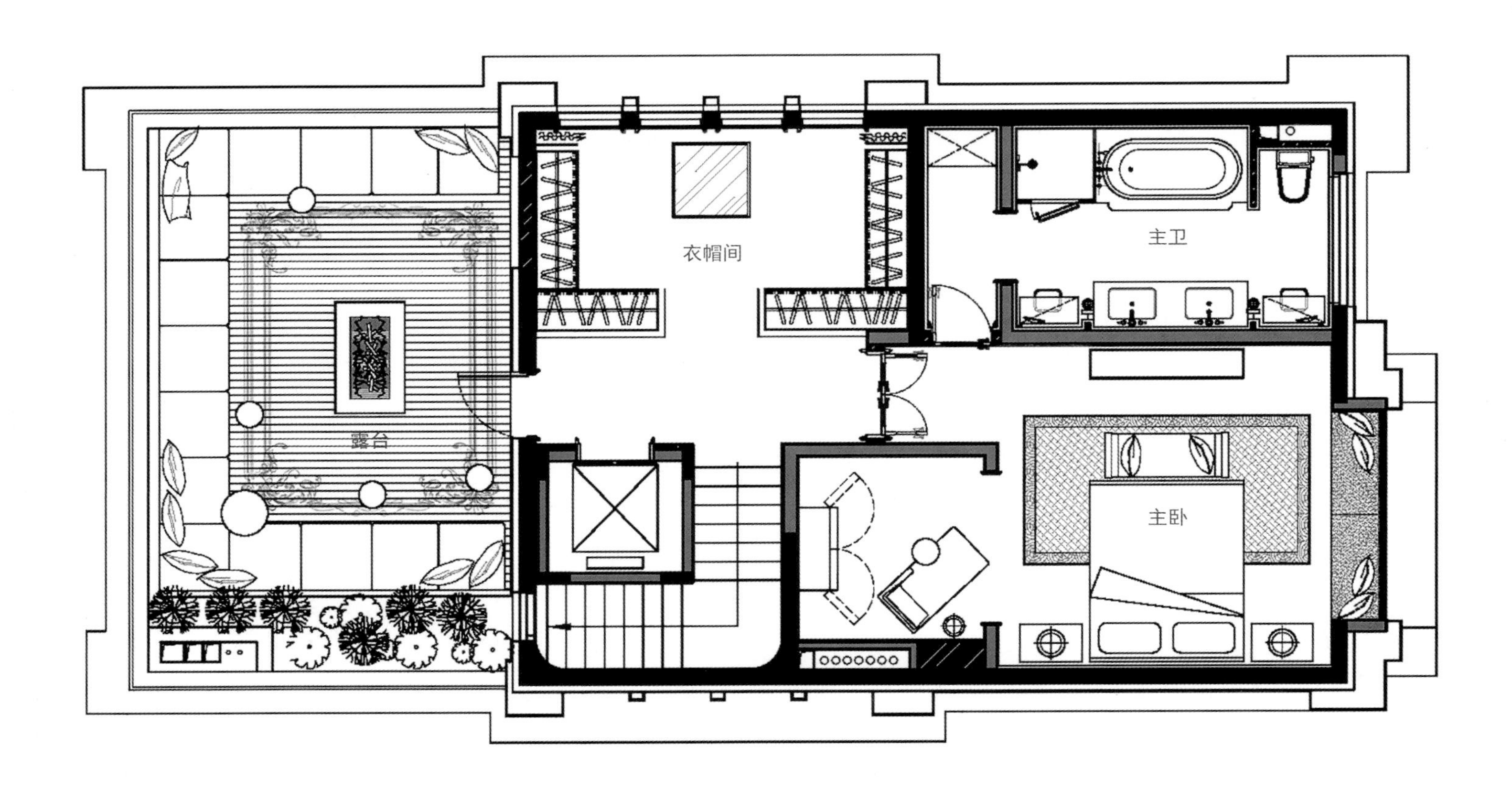
衣帽间
主卫
露台
主卧

整个卧室泛着蓝宝石光泽的灰色，显得尤为奢华，床沿背板包裹的灰色丝绒更是尽显法式的矜贵。除了软装饰品，主卧还考虑到主人的便捷和需求，增加了一处家庭 Mini Bar 的功能区，提升了空间的使用品质，同时也能服务到北侧 36 平方米的私有露台。卫生间的布局同样使用对称布局的手法，体现空间整齐流畅的使用动线，双台盆、双梳理台与台盆呼应了独立浴缸、淋浴区及马桶间，主卫出入的廊道更是设置了具有收纳功能的区域。露台以围合形式布置，中心区域安置了火炉，为露台平添了另一种惬意的生活，户外地面的木地板，选用了具有手工描绘欧式纹样的材料，别具一格。独立通透的衣帽间区域有着两个金属框架的玻璃大橱，犹如时装品牌的旗舰店铺，该设计以建筑窗做参照，形成对称布局的高柜衣帽收纳区域及中心首饰收纳柜，体现了主人的生活品质，展现出衣帽间奢华的一面。

地下一层为 137.63 平方米，设有雪茄吧及贯穿地下二层的挑空书房空间，同时也安置了别墅的储藏空间和保姆房。通过雪茄吧进入挑空的书房，那里巧妙糅入图书馆的畅想，两层高的开敞书柜，中间以钢结构的走道上下分隔，沿走道移动可品味不同角度的空间感受。错落有致的书柜与装饰画的“书柜”穿插设置，让人有着“原来如此”的欣喜。书柜的东侧设计了一个机关，推动书柜，内部便是一处暗室，可作为主人私人的收藏区域。

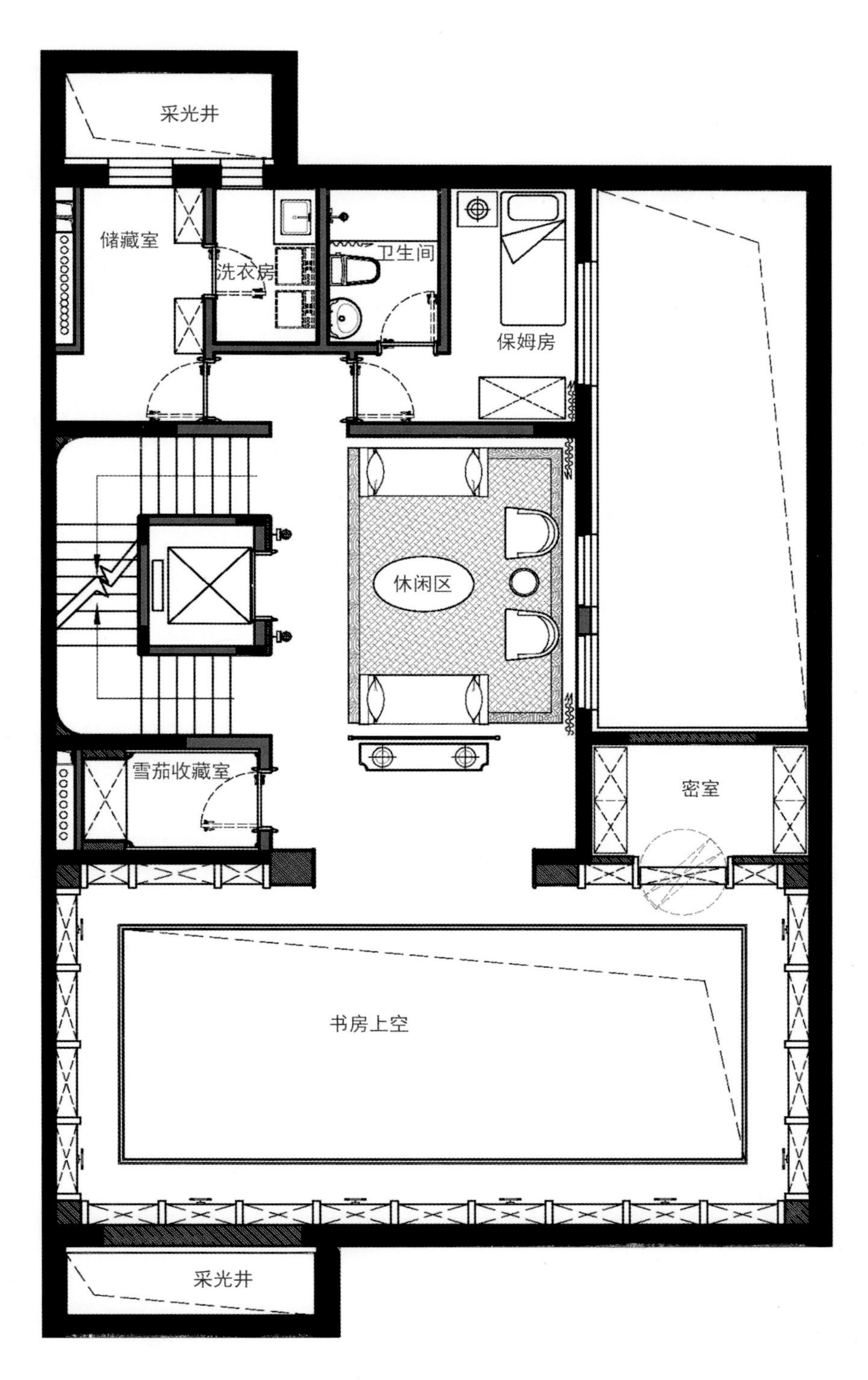

地下二层共有 183.62 平方米，除了挑空的书房，还有家庭娱乐室及相应的服务空间。书房空间的软装饰品及家具突出了男主人的性格，皮质的沙发、独具质感的风化木书桌，开敞大气的书房即刻呈现于人前。书房的东侧端头与客厅采用同样的设计手法，设置了一座手工壁炉。壁炉北侧为下沉的庭院，可通过庭院的楼梯到达一层户外空间，南侧通往车库。设计师在车库与入户的中间区域安置了一处廊道，以收纳功能为主，形成了入户的玄关。

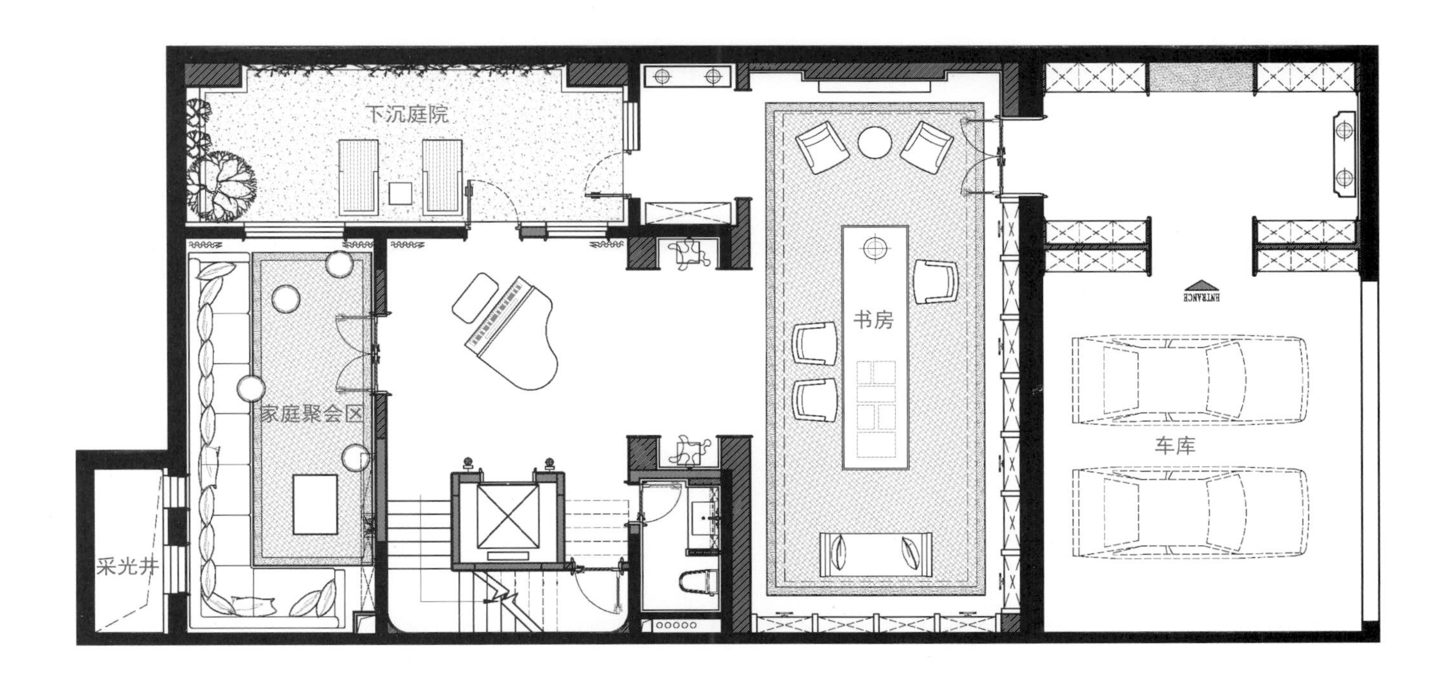
下沉庭院
家庭聚会区
书房
车库
采光井
ENTRANCE

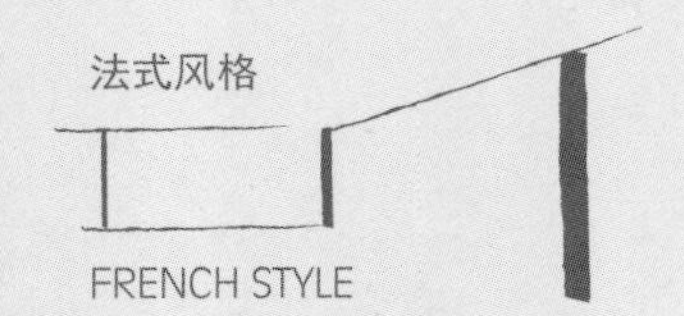

DESIGN CONCEPT 设计理念

This villa traces to the source of traditional European noble cultural temperament, and integrates Lingnan landscape culture, modern lifestyle and international fashion taste. It upholds modern French style and adopts semi-enclosed architectural planning, which offers the interior enough lights and provides beautiful and natural courtyard landscape on three sides. The ostentatious space of five layers combines the void and the solid and promotes progressively through the well-organized space planning. In the blue-green tone of rich layers, the texture silk fabrics collocate with solid wood furniture with black and gold edging. A simple move or action can feel the exquisite and elegant noble charm.

金地天河公馆别墅溯源欧洲传统贵族文化气质，融合岭南山水文化、现代生活方式与国际时尚品味。该别墅秉承现代法式设计风格，采取的半围合式建筑规划，使得室内不仅光照充足，同时三面可享庭院自然景观。纵向五层的阔绰空间，透过主次有序的空间规划，虚实相间，层层推进。在层次丰富的蓝绿色调中，极富质感的丝绸布艺搭配黑金描边的实木家居，俯仰皆能感受到精致优雅的贵族韵致。

FRENCH NOBLE AND ELEGANT RESIDENCE

法式贵族雅宅

项目名称：天河公馆	设计公司：深圳市则灵文化艺术有限公司	设计师：罗玉立
项目地点：广东广州	项目面积：600 平方米	摄影师：黄书颖

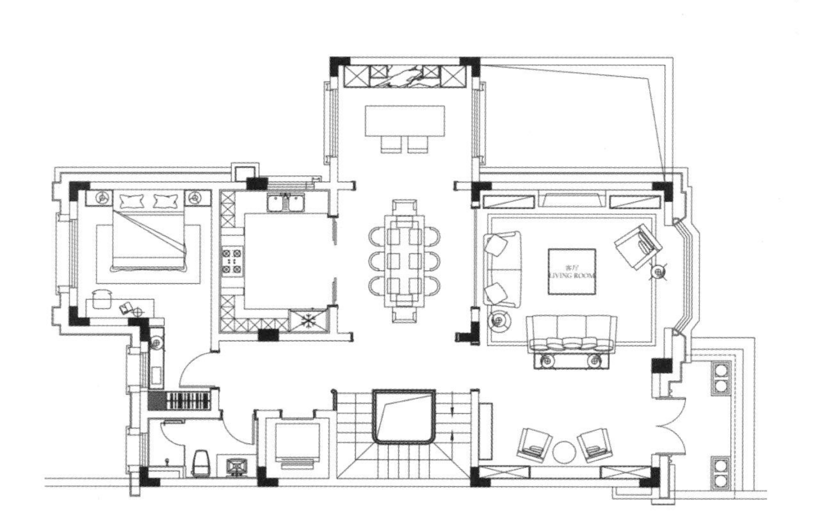
客厅
LIVING ROOM

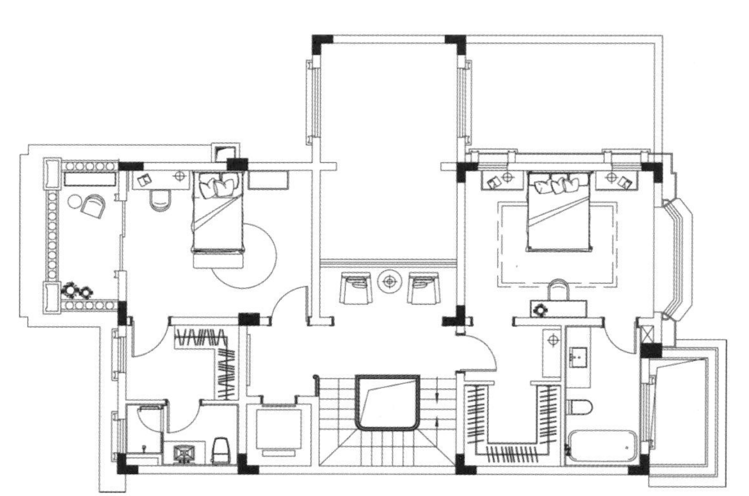

Self-Portrait U.S.A.

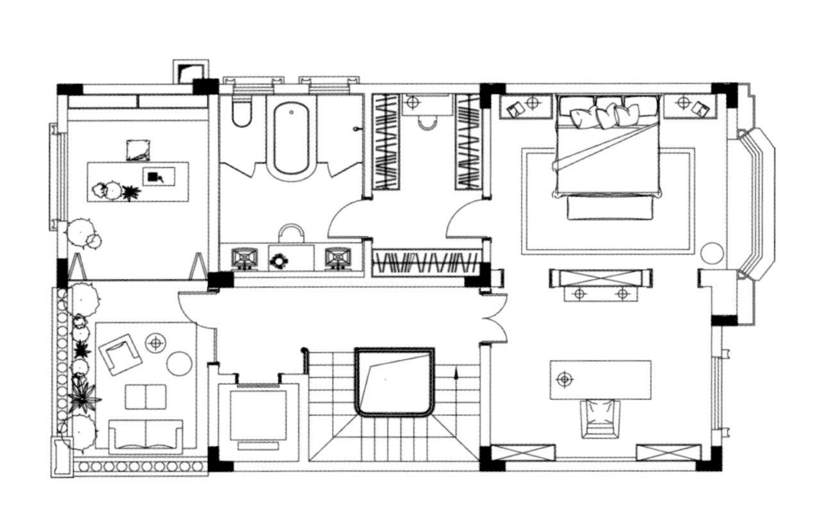

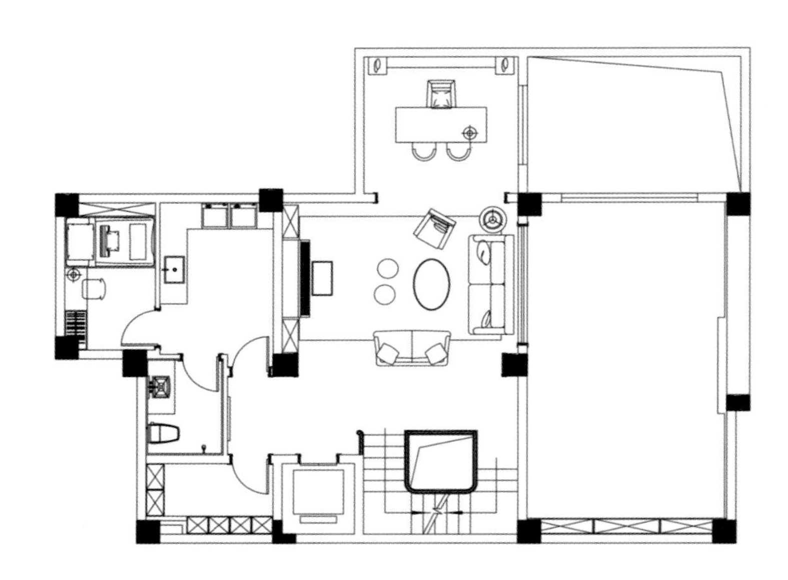

ROCK
LOS BEATLES

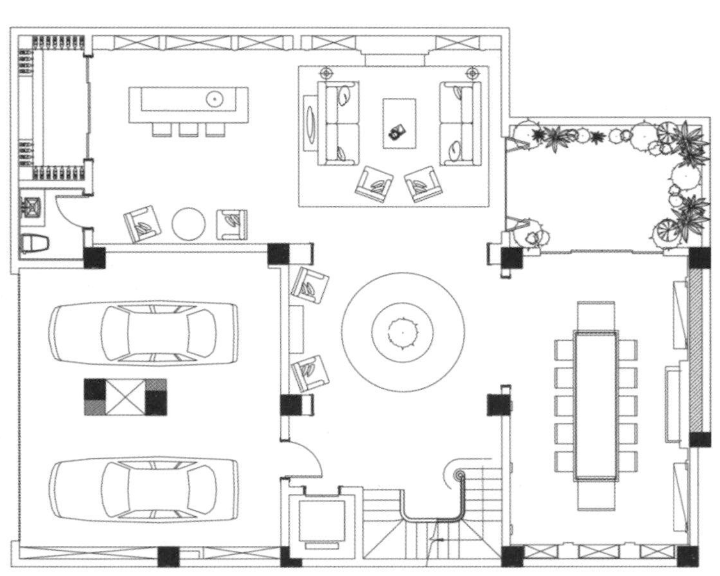

DESIGN CONCEPT 设计理念

The designer uses Chinese and European techniques to conduct the design. The shapes of furniture consider the actual scale of the space and avoid the problem of large size European furniture. The designs refine cultural symbols from two kinds of decorative styles, integrate Chinese elements with European style and create an artistic living atmosphere, which collides unique artistic sparkles in the Chinese and European space.

设计师采用中式与欧式混搭的手法进行设计，家具的选型充分考虑了实际空间的尺度，避免欧式家具体量过大的问题。设计提炼了两种装饰风格的文化符号，将中式元素与欧式风格相互融合，打造一个具有艺术特色的家居环境，中欧混搭的家居空间碰撞出更为独特的艺术火花。

ORIENTAL SENTIMENT AND WESTERN CHARM

东情西韵

项目名称：苏州海亮·唐宁府	设计公司：GND 设计集团——N+ 恩嘉陈设	设计师：宁睿
项目地点：江苏苏州	项目面积：320 平方米	摄影师：大斌
主要材料：布艺面料、水晶灯、软包等		

Corporate Design
NEW CITY PARKS

The mix-up design of Chinese style and European classical style endows the space with distinct fun. Luxurious gold intersperses the space and the linear modeling breaks the regular and dull atmosphere. The wallpaper of the master bedroom is the highlight of the space, which adds heavy decorative effects to the exquisite house. Luxurious soft bed back and copper desk lump present the taste and fashion of the owner. The repeated elements of the wall strengthen the impression. The mirror decoration unifies the tone of the space and extends it. The organized arrangements with exclusive rhythm maintain the balance of the space. The unique lines of the mirror in the bedroom add beauty to the space. The mirror and even the whole house are divided into two space by the lines which lead the sights. The ubiquitous details such as flowers, music, presbyopic glasses and a book can present the tranquil later life of the elder. The tone of the children's room is blue and white, which achieves unrestrained and far-ranging fantasy of the children. The desk echoes with the bed, which strengthens visual balance. The theme of baseball highlights the characteristics of boy. The big baseball ornament lightens the whole space.

中式风格与欧式古典风格的混搭设计，使得空间碰撞出别样的趣味。奢华的金色点缀空间，流线造型打破规矩、呆板的氛围。主卧壁纸的选择可为亮点，为精装修的住宅增添了厚重的装修效果。奢华的软包床背、全铜的台灯，无处不体现着主人的品味与时尚。墙面重复的元素加强给人的印象，镜面装饰不仅使画面统一还让空间得到延伸，有规律的排列有着专属的节奏感，维持着空间的平衡。次卧镜子独特的线条为空间增添异彩，镜子乃至整个房间像被线条分割成两个空间，引导着视线。鲜花、音乐、花镜与一本书，无处不在的细节刻画，体现着老人恬淡的晚年生活。儿童房的设计以蓝、白色为主，满足了孩子海阔天空的幻想。书桌和床遥相呼应，增强了视觉的平衡感。棒球主题突出了男孩特征，超大棒球挂饰为整个空间点燃亮点。

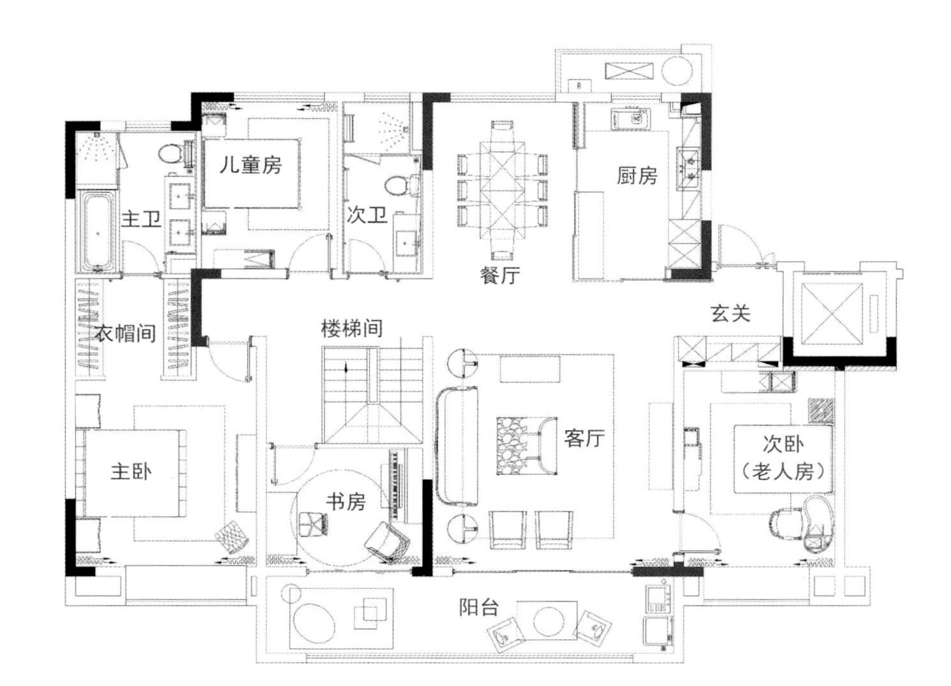

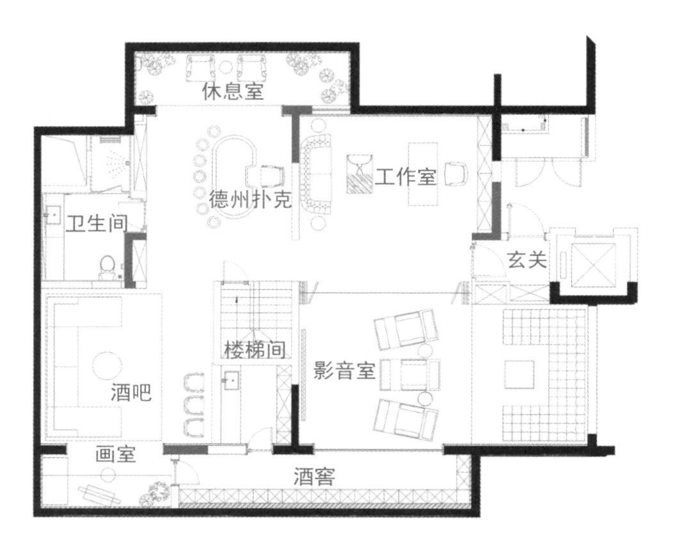

The high-quality and exquisite decorations present the unique high taste. The basement contains a large audio-video room, a red wine and cigar bar, a wine cellar, a Texas poker area, a large study, a reception area and a skylight studio. With multilayer house products and enjoyment of a villa, the basement is designed according to the man's social activities and makes a contrast with the warm living atmosphere upstairs, with natural color background interspersed with noble blue.

高品质精致的饰品呈现独一无二的高品位。地下室囊括了超大影音室、红酒雪茄吧、酒窖、德州扑克区、超大书房及接待区、天光画室。多层洋房的产品，别墅级的享受，地下室的设计以男主人的社交为主线进行设计，与楼上温馨的居家氛围形成对比，以自然色调为底点缀以高贵的蓝色。

YOUNG CHANG

DESIGN CONCEPT 设计理念

The collocation of glorious and bright colors, such as gold, yellow and red, renders extraordinary temperament of the European palace. The ceilings and wall ornaments with carved flowers and gold and the classical furnishings with colorful paintings and gold ornaments convey the elegant aesthetic taste and aristocratic cultural connotative life attitude of the owner by a graceful and sumptuous posture. The design of this show flat is not limited by thoughts and models. The custom-made luxurious space brings elegant and exalted feelings in the European palace, which achieves the noble palace dreams of city elites.

金、黄、红等辉煌明亮的色彩搭配，渲染出欧式宫廷不凡的气度。镶花刻金的天花墙饰与彩绘金饰的古典陈设，均以一种雍容、华贵的姿态，传递着居室主人高雅的审美情趣和极富贵族文化底蕴的生活态度。该样板房设计没有拘泥于思维的禁锢和样板的束缚，量身定制的奢华空间，带来欧式宫廷优雅尊贵的生活感受，一圆城市新贵的宫廷贵族梦。

PALACE DREAMS OF CITY ELITES

城市新贵宫廷梦

项目名称：南宁霖峰壹号欧式宫廷风格样板房设计	设计公司：KSL 设计事务所	设计师：林冠成
项目地点：广西南宁	项目面积：130 平方米	
主要材料：闪电米黄大理石、黑檀木纹大理石、拉丝古铜、木饰面、墙纸、银镜等		

Entering the living room, magnificent beige marble floors, ceilings and wall ornaments with carved flowers and gold, exquisite classic furnishings and beautiful palace paintings bring you into a classical world. Graceful crystal droplight and exquisite silver mirror add beauty to each other. The dense European palace style presents an extraordinary atmosphere. After designer's delicate treatment, the Ariston fireplace with exquisite carvings keeps the classical European living feelings and skillfully reconciles the needs of modern life. Every corner in the dining room presents noble, elegant and magnificent feelings. The designer injects artistic inspirations into daily life by diligent designs. What you enjoy is absolutely an improvisation which manifests noble taste and modern fashion.

走进客厅，富丽的米黄大理石地面、镶花刻金的天花墙饰、考究的经典陈设和精美的宫廷油画瞬间将人引入古典世界，令人目不暇接。华美水晶吊灯与精致银镜交相辉映下，浓郁的欧式宫廷风情，显露出令人凝神屏息的不凡气场。纹饰镌刻精美的雅士白壁炉，经过设计师匠心独运的处理，既保留了古典欧式的生活情怀，又巧妙地调和了现代生活的需求。餐厅空间以贵族优雅瑰丽的情愫潜入每一个角落。设计师以美的笔触，将艺术的灵感注入日常生活中，你所享受的，绝对是一次贵族品位与现代风尚相得益彰的即兴创作。

When soft and elegant noble temperament combines with comfort and warmth of modern life, one can ponder the pace of European palace and city elites again to seek for lives of their own. The bright crystal droplight, fancy palace painting, luxurious bedding and classical color collocation in the master bedroom can bring people endless noble daydreams.

当柔和高雅的贵族气质牵手现代生活的舒适温馨，重新揣摩欧式宫廷与城市新贵的节奏，找寻最属于自己的生命轨迹。主卧室璀璨的水晶吊灯，静美的宫廷油画，奢华的床品以及古典的色彩搭配，你可以想象这样的家将给人带来怎样的贵族遐想……

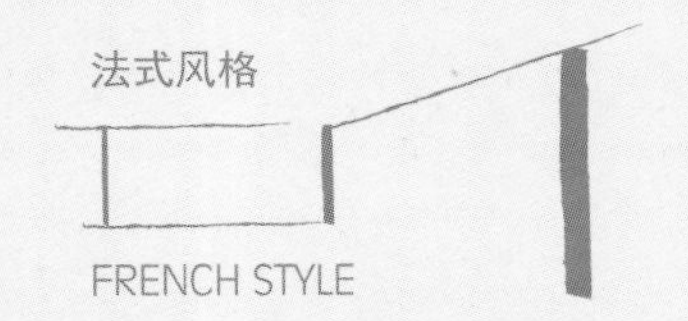

DESIGN CONCEPT 设计理念

Shanghai is never lack of prosperity and dignity. But if you never contact it, you cannot image it forever. To those who are in the top of the pyramid, "best" is neither high prices nor piles of expensive products. It is the calmness and elegance which can see the distance. This project exactly prospects the new trend of world mansion lifestyle and fills up the blank of it near Huangpu River. The unprecedented land of idyllic beauty in city core is born.

Starting from the needs of life feelings and living comfort, the designers aim to make every resident be flexible between career and home, luxury and tranquility, which casually manifests upper class temperament and low-key luxury in the static and dynamic.

上海，从来不缺繁华和尊贵，但你若不曾接触，便永远无法想象。对身处其中的这部分金字塔尖人群来说，“最好”不再仅仅是数不清零的价格，也不是昂贵产品的堆积，而是可以望见远方的从容和优雅。融创滨江壹号院正是前瞻了世界豪宅生活方式新趋向，在黄埔江畔填补了这种生活方式上的空白，前所未有的都市核心世外桃源意境就此而生。

从生活感受和居住舒适度的需求出发，设计师旨在让每一居住于此的业主都能在事业与家庭、繁华与静谧之间游刃有余，在动静之间不经意显露上流气质和低调奢华。

AESTHETIC STATE
唯美之境

项目名称：融创滨江壹号院 D 户型	设计公司：上海益善堂装饰设计有限公司	设计师：王利贤、张琳琳、宋莹
项目地点：上海	项目面积：223 平方米	摄影师：温蔚汉
主要材料：大理石、水晶、布艺等		

The noble temperament emits rural feelings. The luxurious and complicated designs are relaxing and tranquil. Most part of the living room is under tall structure. Large areas of French windows bring good day lighting, which makes the vision of the whole space wider. The panoramic view of Huangpu River is seen, which is magnificent. The fabric sofas have velvet texture and fluent wood curves. The unified American dark color hardwood tables and desks match with exquisite carves, gold plating and master slice, which constitutes many beautiful furnishings without any heavy and rough sense.

高贵的气质中散发着些许乡村情怀，奢华繁复的设计却不失轻松恬静。客厅的大部分在挑空结构之下，大面积的落地窗带来了良好的采光，让整个空间的视野更加开阔，黄浦江的美景尽收眼底，很是气派。布艺沙发组合有着丝绒质感以及流畅的木质曲线。统一的美式深色硬木桌柜，与精细的雕刻、镀金及贝母片的融合下俨然一件件精美的装饰品，没有留下一丝一毫的厚重与拙朴。

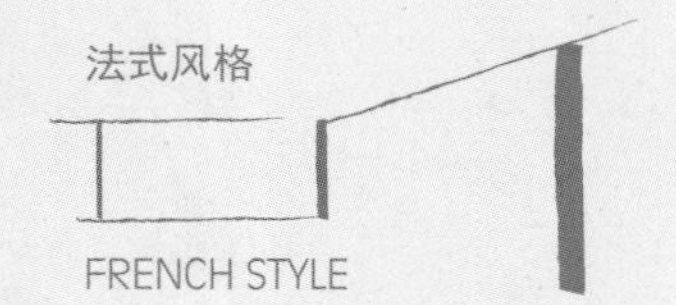

DESIGN CONCEPT 设计理念

The beautiful Seine
You are a noble gentleman
And a tender and romantic lover.
The breeze kisses your cheek
The cloud clusters your tenderness.
The leaves dance for you
The drizzle is in love with you.
How many touching stories do you collect
How many figures of lovers do you capture?
Hold your hands
In the back garden of Chateau de Versailles
Our romantic footprints are left.

Appointment—To the Seine

ROMANCE NOT IN PARIS

不在巴黎的浪漫

项目名称：成都 · 中洲中央公园蓝湖郡 T3 别墅
设计公司：深圳市逸尚东方室内设计有限公司
软装设计：江磊 、林叶
项目地点：四川成都
项目面积：303 平方米
主要材料：卡金灰大理石、爵士白大理石、帕斯高蓝大理石、北极光大理石、胡桃木木地板、墙纸、橡木木饰面、皮革等

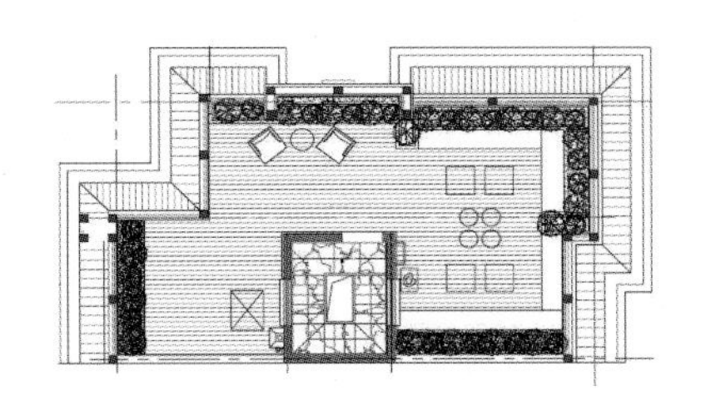

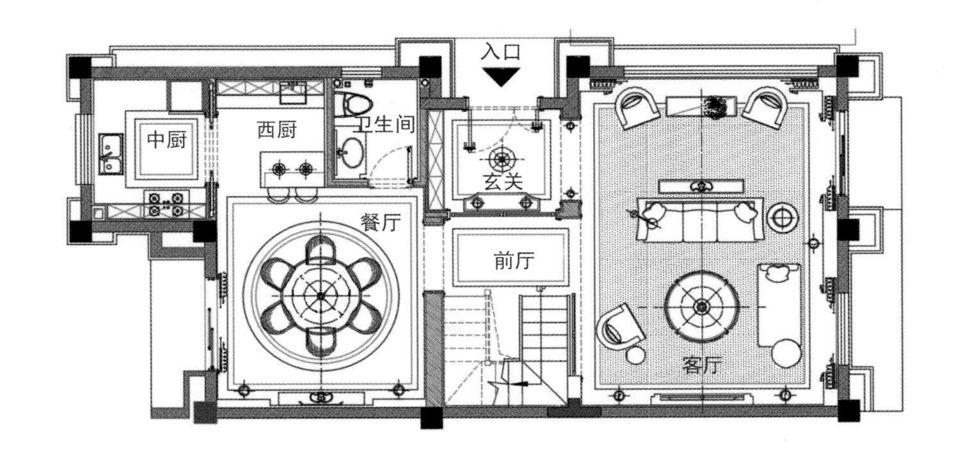

美丽的塞纳河
你是一位高贵的绅士
又是温柔浪漫的大众情人
微风亲吻你的脸颊
白云簇拥你的柔情
树叶为你翩翩起舞
细雨为你动了真情
你搜集了多少动人的故事
你捕捉了多少恋人的身影
与你牵手
凡尔赛宫的后花园
留下了我们浪漫的足迹
《约定——致塞纳河》

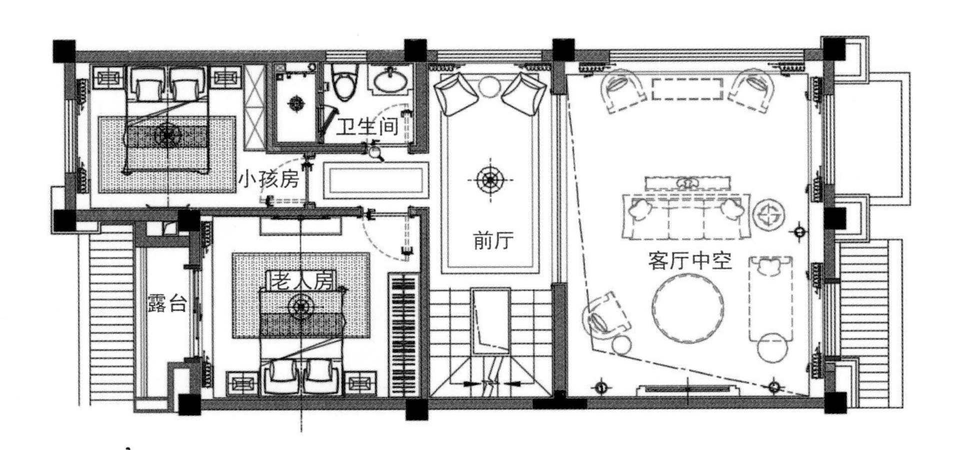

The designers vividly deduce the foreign lifestyle with “exceedingly fascinating and charming romantic feelings in the Seine”, penetrate the elegant, natural and artistic French noble flavor into every corner, strengthen historic and cultural taste by innovative interior furnishings, and relate the exquisite and delicate life of people in the villa. Luxury to the heart is the extreme luxury, and enjoyment to the thought is the supreme enjoyment.

Colors stimulate dreams, and dreams change lives. Starting from here, a space, a story, accessories mixed in various styles...

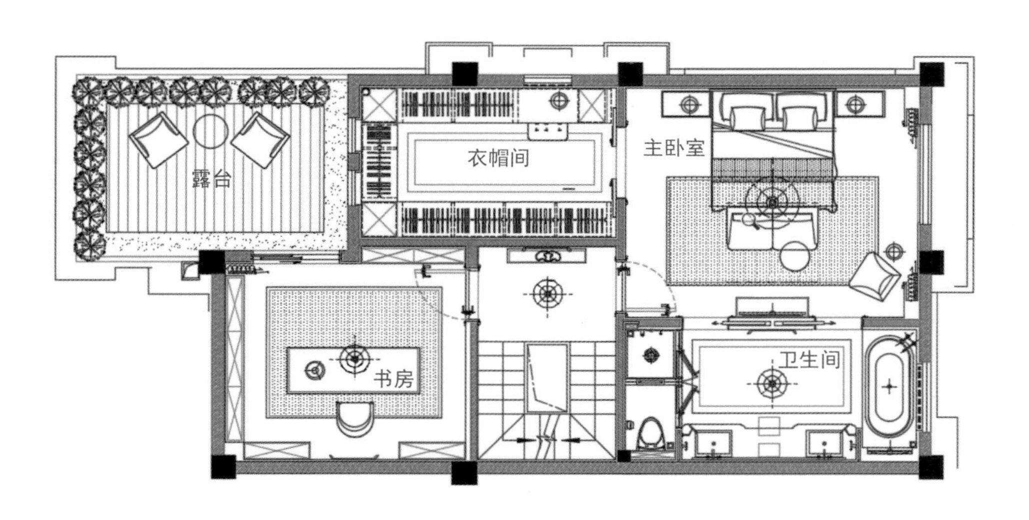

设计师形象地演绎“风情万种的塞纳河浪漫情怀”的异域生活方式，将优雅的、自然的、艺术的法式贵族气息渗透在每个角落，用创新的室内陈设来增强历史与文化气息，将人们在别墅内的生活细腻而又精致娓娓道来。奢华于心才是极致的奢华，享乐于思才是至高的享乐。

色彩激发梦想，梦想改变生活。从这里开始，一个空间一个故事，多种风格的混合配饰……

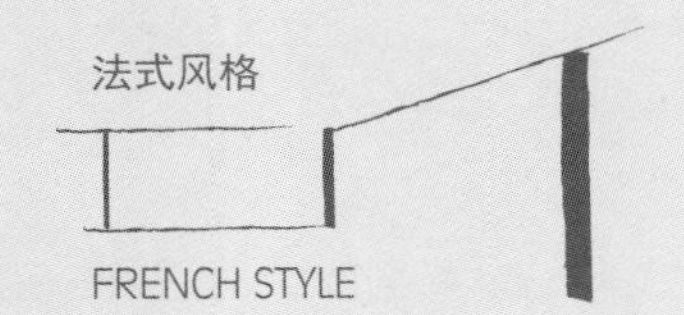

DESIGN CONCEPT 设计理念

French people are never lack of romance. They believe in the aphorism by Balzac that elegant life is the free play of the mind. And there is a Chinese verse "the pearl white is penetrated into the misty rain, and the peacock blue reflects the light of the moon". When Chinese red meets with French blue, it creates a distinct beauty of life and displays a tranquil and romantic French painting for the villa.

法国人从来就不缺乏浪漫，他们坚信巴尔扎克的格言：风雅生活是心灵的自由发挥。中国的诗句"珍珠白沁就烟雨，孔雀蓝映著月光"。当中国红遇见法国蓝，碰撞出的是别样的生活之美，为我们这个别墅呈现出一幅恬淡、浪漫的法式画卷。

POETRY FROM THE EXTRAORDINARY TEMPERAMENT

非凡气质中的诗情画意

项目名称：杭州绿城桃源小镇荣景苑法式别墅

设计公司：西象建筑设计工程（上海）有限公司

设计师：何文哲

项目地点：浙江杭州

主要材料：大理石、吊灯、玻璃、木质地板等

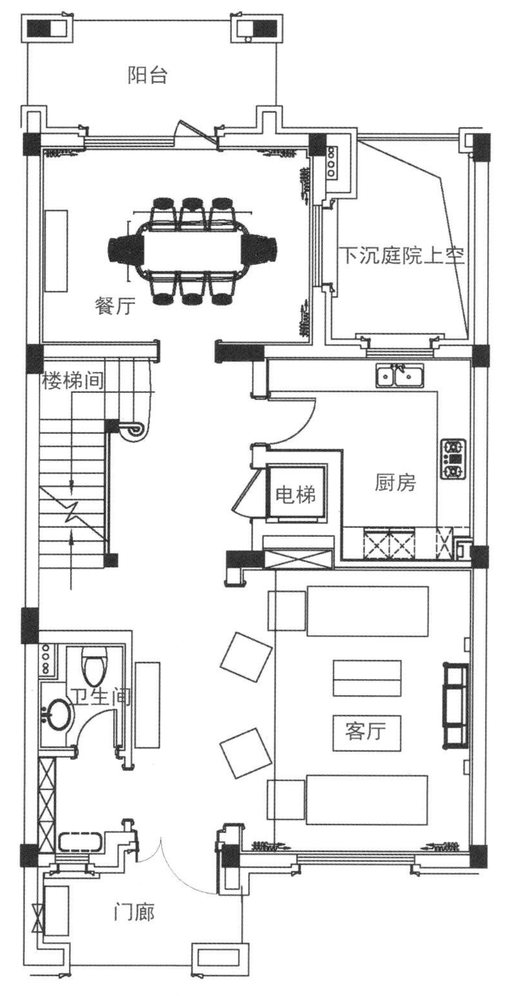
阳台
下沉庭院上空
餐厅
楼梯间
电梯
厨房
卫生间
客厅
门廊

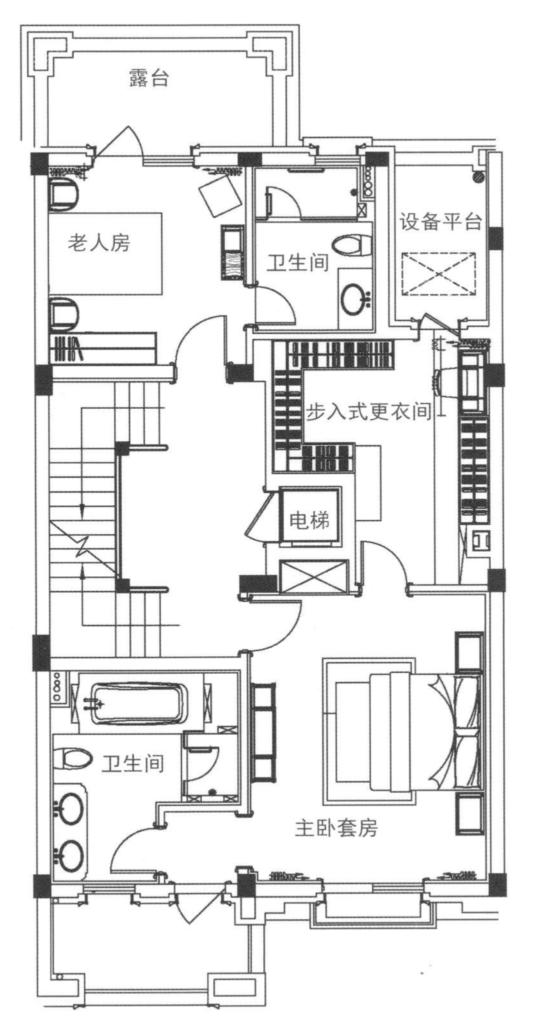
露台
老人房
卫生间
设备平台
步入式更衣间
电梯
卫生间
主卧套房

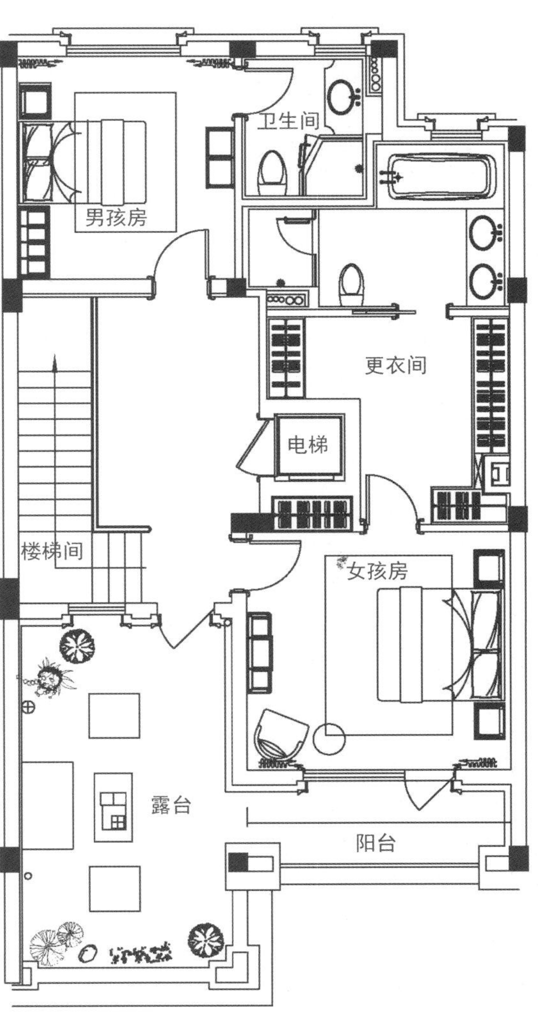
卫生间
男孩房
更衣间
电梯
楼梯间
女孩房
露台
阳台

The furniture in the living room is elegant and luxurious with enthusiastic and gorgeous colors. The blue and white porcelains present Oriental beauty. The main tone of the dining room is elegant blackish green. The bone china tableware and crystal candlestick create the lifestyle of French nobility in 18th century. The layout of the staircase is spacious. Go upstairs, you can see the master bedroom suite. Continuing the elegant and gorgeous colors in first floor, the main tone of master bedroom is dark blue. It abandons complicated crafts, presents elegant curves of the furniture, and combines traditional French blue, orange, silk and embroidery to create heavy Oriental colors. Every turn-back is a gorgeous scenery. Every detail manifests dense French style. The main tone of the basement is blue gray, white and red, interspersed with emerald and orange, which creates comfortable and relaxing leisure atmosphere and reflects fortitudinous personality of the man of the house.

客厅家具优雅华丽，色彩热烈绚烂，饰品以青花瓷点缀，呈现东方之美。餐厅色调以优雅的墨绿为主，骨瓷餐具与水晶烛台觥杯交错，营造了十八世纪法国贵族的生活方式。楼梯间布局宽敞，上楼即可看到主卧大套间。沿用一层优雅华丽的色彩，主卧以深蓝色为主调，摒弃了繁琐的工艺，展现家具本身的优雅曲线，并结合传统法式的蓝色、橘色，以及丝质、刺绣面料营造了浓烈的东方色彩。每个转身都是一道华丽的风景，每一处细节都彰显了浓浓的法式情。地下室空间以蓝灰色、白色和红色为主色调，点缀翠绿色和橙色，营造舒适轻松的休闲氛围，也体现了男主人坚毅的性格。

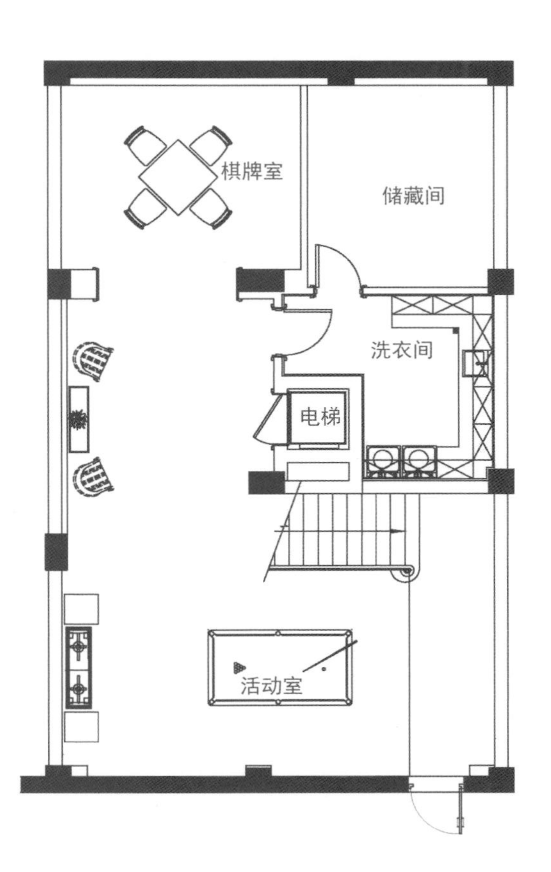
棋牌室
储藏间
洗衣间
电梯
活动室

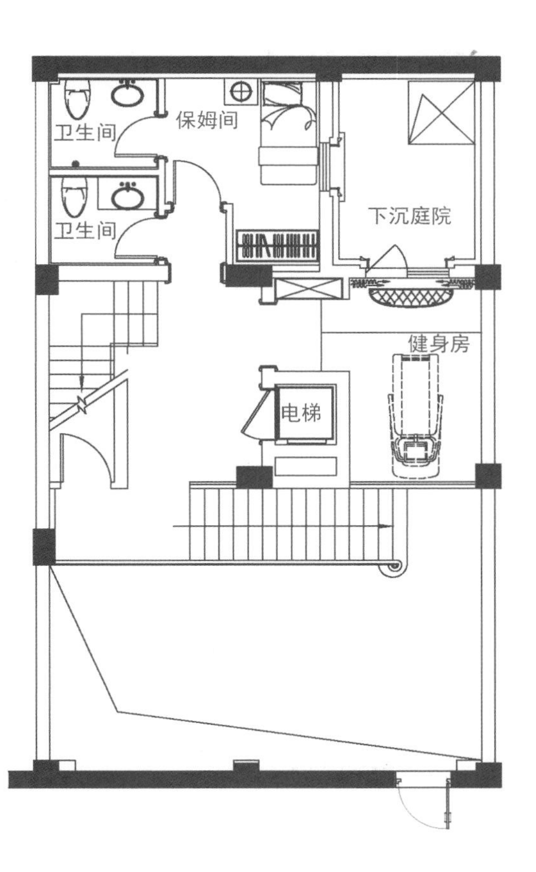
卫生间
保姆间
卫生间
下沉庭院
健身房
电梯

简欧风格

SIMPLE EUROPEAN STYLE

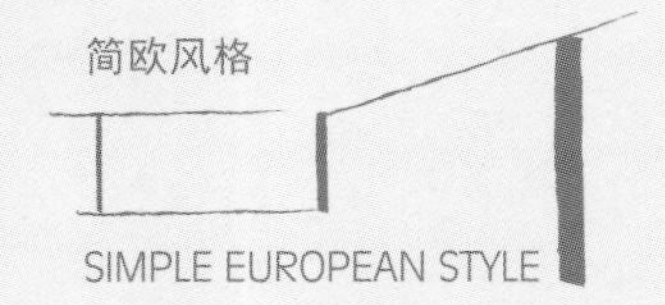

DESIGN CONCEPT 设计理念

Located at Lion Mountain of one thousand years, this show flat, as a masterpiece of Longfor Properties in Suzhou and based on 24 years villa creation experience, is decorated by Zest Art to salute to "scholar" culture in Suzhou by an authentic, sincere, low-key and restrained attitude. For Suzhou, in the past there was Tianyi Pavilion through a century of wind and rain, and more recently there was Wenlan Pavilion which was connotative and magnificent. Scholars' pursuit of book collections comes from the nation's expectations of forever civilization. "The good influence of men of virtue will not last more than five generations." Book collection to later generations is a century foundation of the family. The underground library is set in the 5.7 meters high villa with daylight above and ventilation below. The ingenious designer abandons the complicated space display and disturbance, only keeps the purest books, the most concise thoughts and innocent belief, and sprinkles a wall of green plants to endow the space with abundant vitality.

SUZHOU SCHOLAR PAVILION

苏州文士阁

项目名称：狮山原著——叠拼	设计公司：深圳市则灵文化艺术有限公司	设计师：罗玉立
项目地点：江苏苏州	项目面积：340 平方米	摄影师：曾康辉、黄书颖

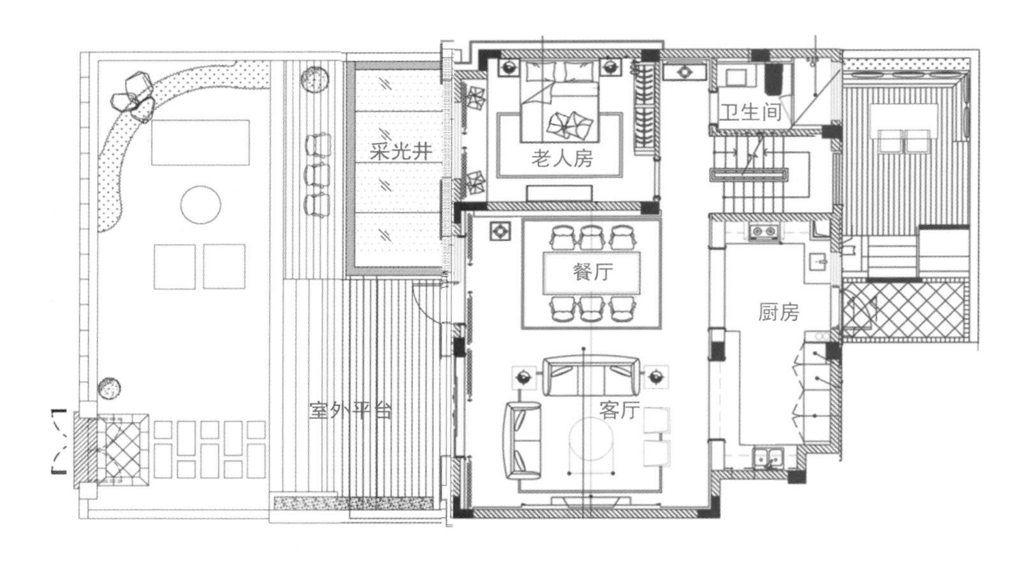

采光井
老人房
卫生间
餐厅
厨房
室外平台
客厅

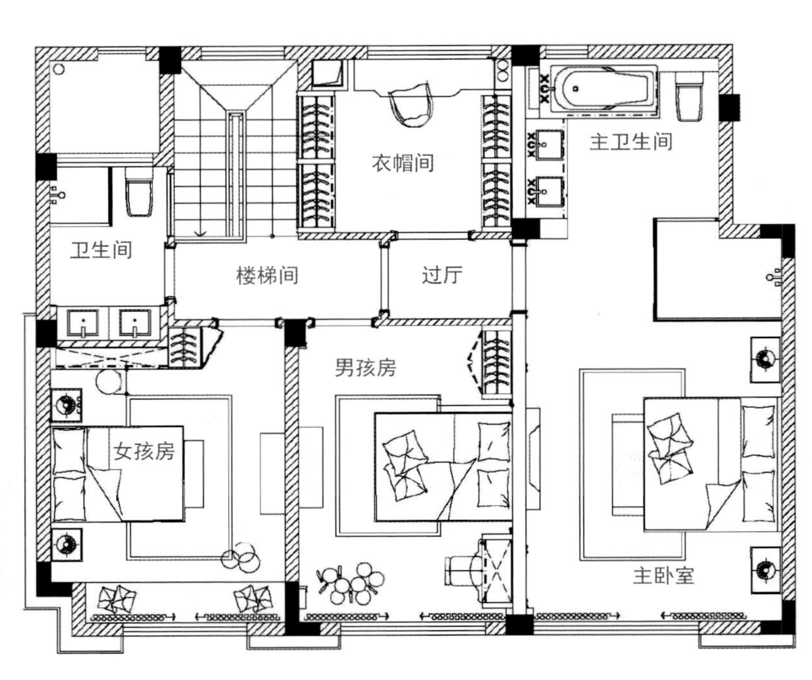
衣帽间
主卫生间
卫生间
楼梯间
过厅
男孩房
女孩房
主卧室

坐落于千年狮子山脉的龙湖顶豪原著样板间，作为龙湖在苏州集 24 年造墅经验的大成之作，由则灵艺术担纲别墅软装陈设设计，以不造作、不浮夸、低调而克制的态度，向苏州“士”文化致敬。于苏州而言，远有天一阁风雨百年，近有文澜阁渊渟岳峙。文人对藏书的追求源于民族对文明火种长存的企盼。“君子之泽，五世而斩”，然而藏书以传后世却是家族百年基业。设置在挑高 5.7 米的别墅地下藏书阁，上有天光引入，下有通风循环。设计师独具匠心，将空间陈设中的繁杂与干扰通通去除，只剩下最纯粹的书、最凝练的思想以及单纯的信仰，再挥洒一墙绿植，点活一室的盎然生机。

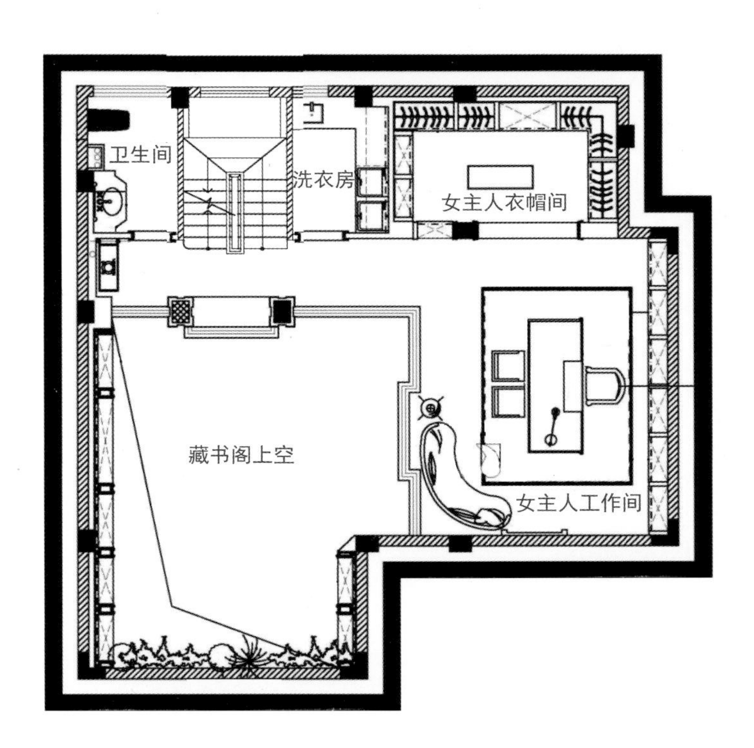

卫生间
洗衣房
女主人衣帽间
藏书阁上空
女主人工作间

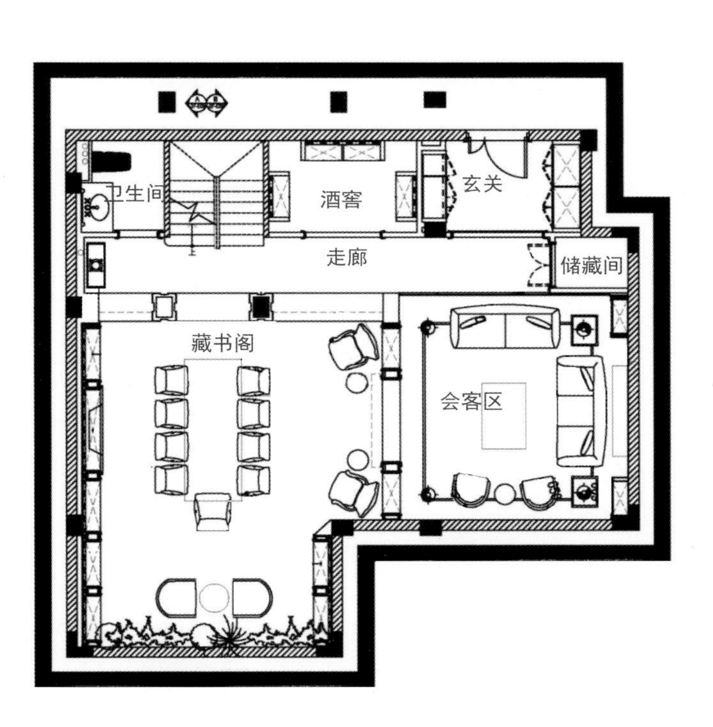

卫生间
酒窖
玄关
走廊
储藏间
藏书阁
会客区

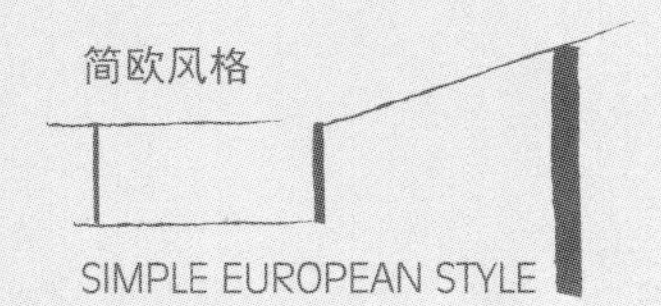

DESIGN CONCEPT 设计理念

The bottom floor is the high-end product promoted by Longfor Lion Mountain. Zest Art is in charge of the soft decoration. Based on French style, the soul with beauty of life endows the space with elegant, fresh and natural atmosphere. With Western style as the soul and Chinese style as the spirit, the designer outlines an inclusive, elegant and interesting space. Let the heart breath freely, and let life return to life. The colors of the whole show flat have great originality. The new willow is nattierblue; the thin string instruments are plain yellow; the background is white. The smoke gathers; the gold is homogeneous; the pool is clean and shallow; what a famous early spring painting, which makes people feel good at first sight. It is as authentic and restrained as a static water flow, which can melt the remaining ice layer of winter.

ELEGANT AND INTERESTING SPACE

雅趣间

项目名称：狮山原著——跃层　　设计公司：深圳市则灵文化艺术有限公司　　设计师：罗玉立

项目地点：江苏苏州　　项目面积：168 平方米　　摄影师：曾康辉、黄书颖

BRÖDMIX
FLERKORN
BRÖDMIX
FLERKORN
SKORPOR

底跃是龙湖狮山原著力推的中高端产品，则灵负责其软装陈设设计，以现代法式做底，将生活之美的灵魂赋予整个空间，淡雅、清新、自成一片天地。以西式为魂，中式为魄，勾勒出一个兼容并蓄的雅趣空间。让心灵自由呼吸，让生活回归自然。整体样板房用色匠心独具，淡青色是新翻杨柳，素黄色是细抹丝簧，白色做底，烟尘初敛，金色调匀，清浅池塘，端景是一幅早春名画，让人初见便如沐春风。不造作，不张扬，如一潭静水深流，融化冬日遗留的冰层。

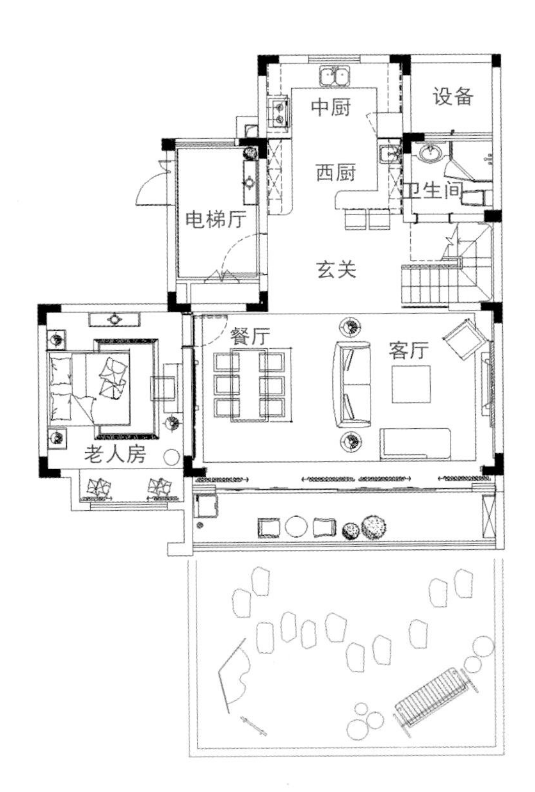

Ballerina

AIR MAIL

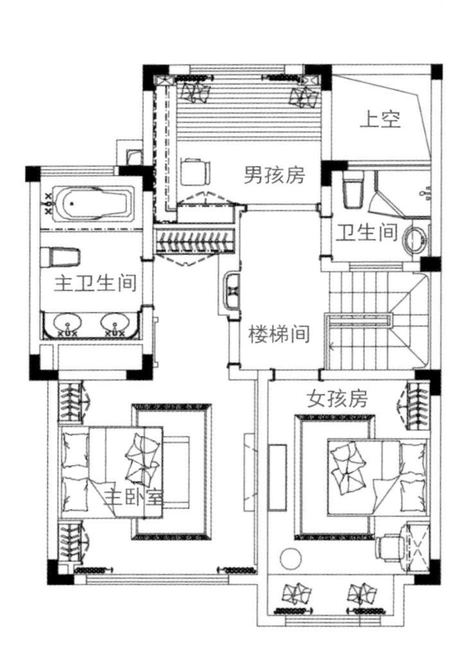
上空
男孩房
卫生间
主卫生间
楼梯间
女孩房
主卧室

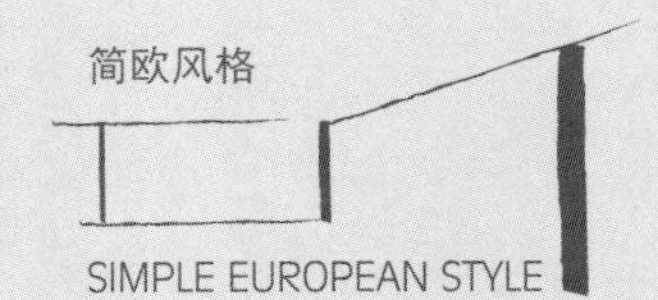

DESIGN CONCEPT 设计理念

The designers use simple materials to build the strength of visual scenes, and use texture materials to pursue a permanent flavor. The clean and concise space logic presents sedate and luxurious temperament. Creating a space with beauty and temperament is not deliberately highlighting the existence of some focus, object or materials, but naturally integrating these elements into the space by a low-key and imperceptible way.

Contemporary life demands complicated space functions so that single function of the previous domain is no longer used and is replaced by closely related new form. The colors of furniture in the foyer, hallway and reception room near the entrance make a sharp contrast, which leaves a deep impression on people. The open kitchen makes the kinetonema of the space more flexible and fluent, which reaches a unified and magnificent effect visually.

GENTLE AND CULTIVATED ARTISTIC STATE

温文尔雅的艺术情境

项目名称：淮安红豆美墅公园里 B1 户型　　设计公司：上海益善堂装饰设计有限公司　　设计师：王利贤、黄亚丽、马玲玲

项目地点：江苏淮安　　项目面积：235 平方米　　摄影师：温蔚汉

主要材料：木饰面、石材、夹丝玻璃、皮革、布艺、不锈钢、墙纸等

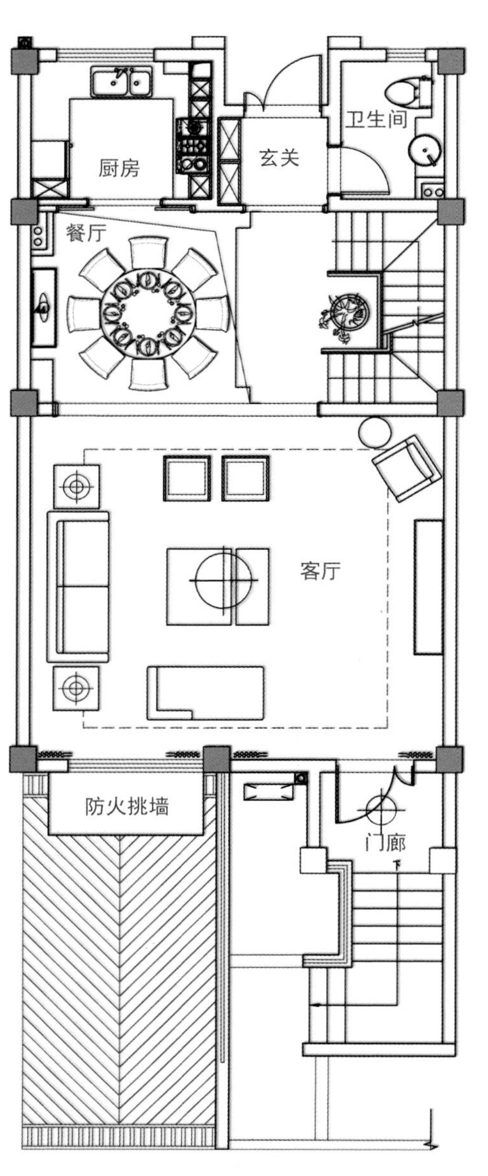

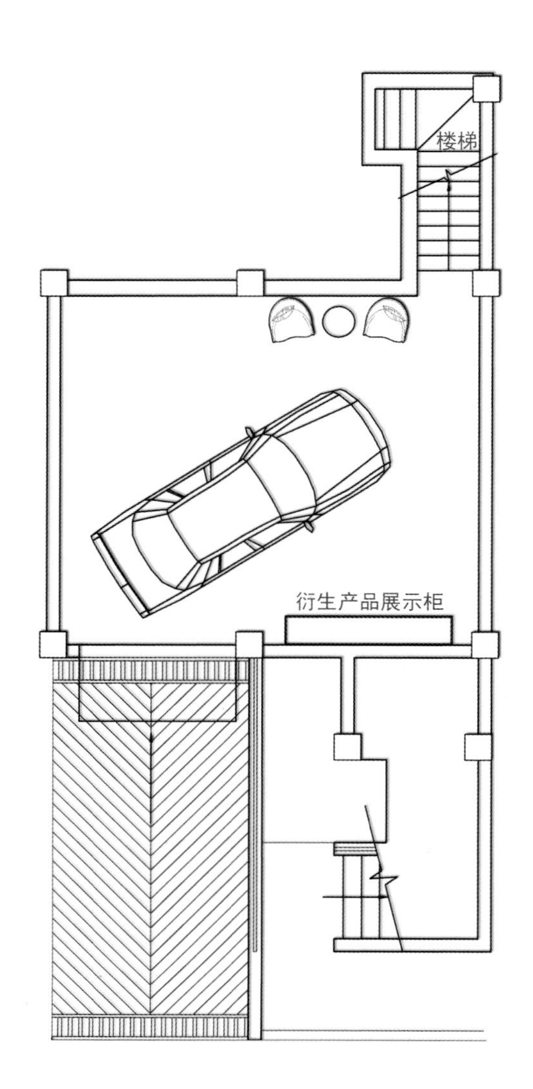

设计师运用品项简单的材料，构建视觉画面的力道，用富于质感的材质，尝试追求一种恒久的情韵。以干净、简单的空间逻辑，来呈现沉稳奢华的气势。构建一种美感与气质真正并存的空间，并非刻意凸显某个焦点、物体或材质的存在，而是以更低调，更让人无法察觉的方式，将这些元素自然糅合于空间中。

当代生活对于空间机能的复杂需求，使得以往场域的单一功能已不再使用，取而代之的是彼此密切相依的新形态。入门处的玄关、门厅和会客室，色彩对比强烈的家具更给人一种眼前一亮的感觉。厨房的开放式处理，令整个空间的动线更加灵活流畅，在视觉上达到统一大气的效果。

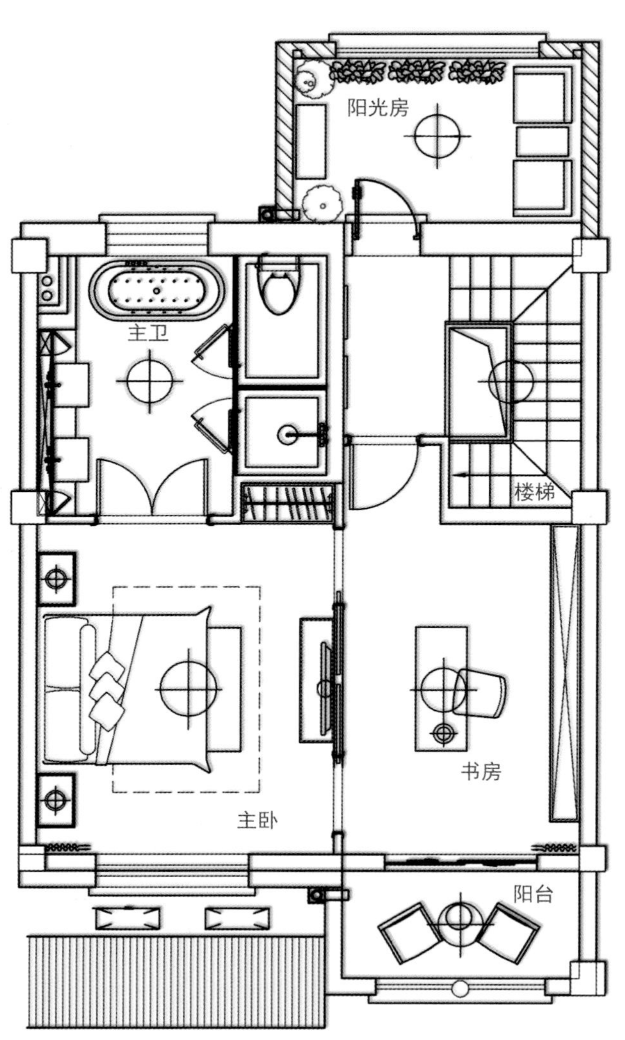
阳光房
主卫
楼梯
主卧
书房
阳台

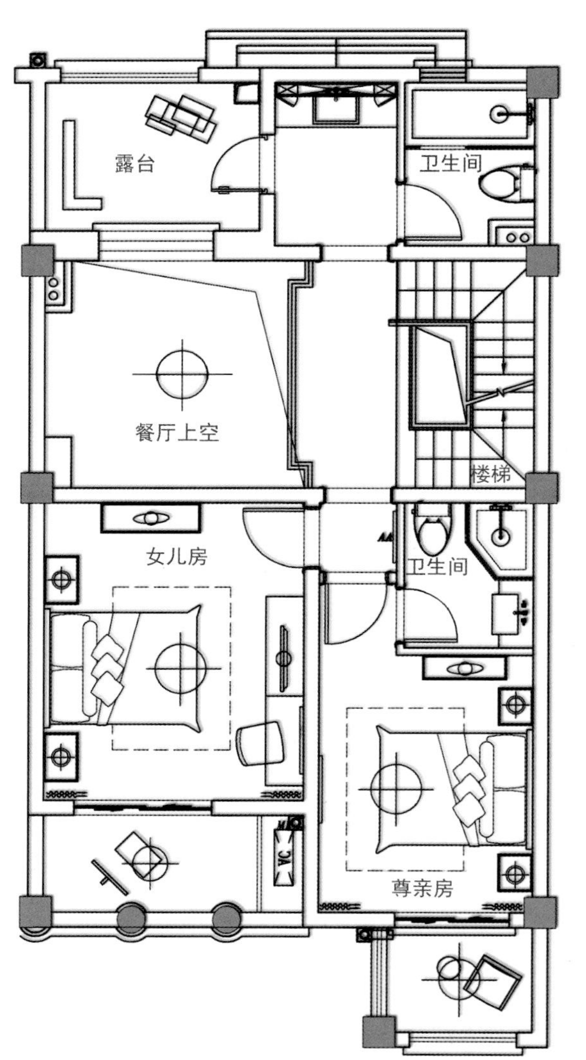
露台
卫生间
餐厅上空
楼梯
女儿房
卫生间
尊亲房

The layout of the interior is coherent with harmonious and unified atmosphere. Natural and exquisite materials and textures collocate with attractive colors, which is a visual enjoyment. Modern design languages combine with texture traditional elements, which endows the space with spirit, sublimates the artistic state of the space and presents gentle and cultivated temperament. Wood veneer encounters with stainless steel, which makes the whole atmosphere clean and well-organized with a connotation “as static as clean pool and as dynamic as ripples”. Living here, people can gain a good rain, nourishment and power, begin a conversation with the soul, and deposit the thickness of life in the never fading times.

室内的空间布局一气呵成，氛围和谐统一，自然与考究的材质肌理与富有感召力的色调搭配，令人赏心悦目。现代设计语言中揉进富有质感的传统元素，赋予空间灵气，升华空间艺术情境，蕴涵温文尔雅的气质。木饰面和不锈钢的邂逅，使整个气氛更显洁净有致，有种“静若清池、动若涟漪”的底蕴。在这般如此的栖息之所，人们获取的是一份甘霖、滋养与力量，开始与自己的灵魂对话，在时光的打凿中永不褪色，沉淀生命的厚度。

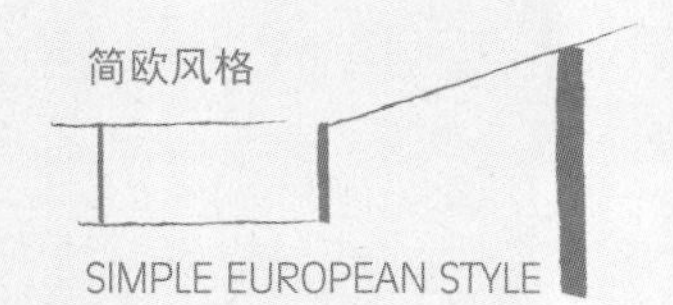

DESIGN CONCEPT 设计理念

This project uses modern European design techniques, absorbs essences from traditional classical style and endows it with new content, and strives to create low-key, luxurious and tasteful high-end residence which provides warm rest place for the president after busy works. The design focuses on fusion of space and natural ecology, and stresses the changes of linear flows. The interior carving crafts concentrate on decoration and furnishings. The colors are gorgeous and use warm colors to coordinate. Transformative straight lines and curves interact, in addition with cat feet furniture and decorative crafts, which creates a magnificent, dignified, concise and modern atmosphere and makes people fully feel the leisure and comfort of modern life.

本案采用了现代欧式的设计手法，汲取了传统古典风格的精髓并赋予了新的内容，力求打造低调奢华、有品位的高端住宅空间，为企业总裁提供繁忙之后的温馨休憩之所。本案构思着重于空间与自然生态的融合，强调线形流动的变化，将室内雕刻工艺集中在装饰和陈设艺术上，色彩华丽且用暖色调加以协调，变形的直线与曲线相互作用以及猫脚家具与装饰工艺手段的运用，构成室内华美厚重而又简约现代的气氛，让人充分感受到了现代生活的休闲与舒适。

MANSION OF THE PRESIDENT

总裁府邸

项目名称：天一国际 ·欧式简约总裁官邸　　设计公司：深圳市山石空间艺术设计有限公司　　设计师：陈岩

项目地点：贵州贵阳　　项目面积：210 平方米

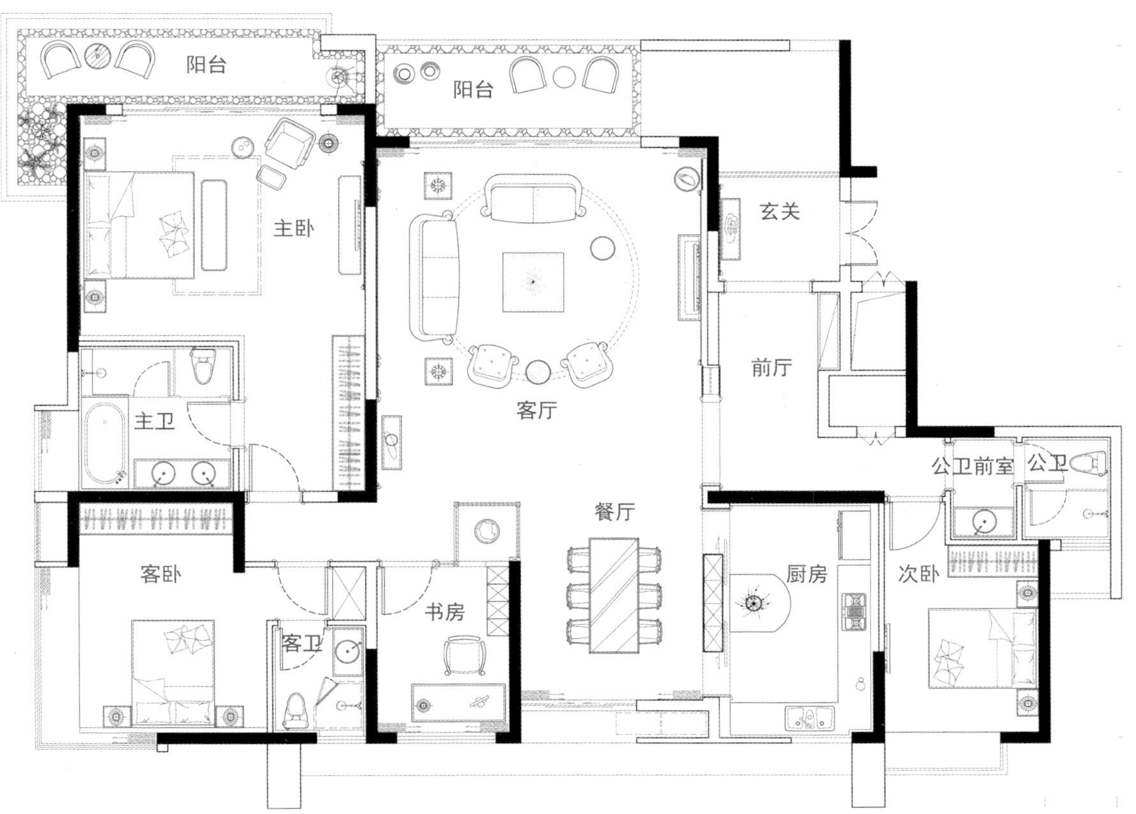
阳台
阳台
主卧
玄关
前厅
客厅
主卫
公卫前室
公卫
餐厅
客卧
厨房
次卧
书房
客卫

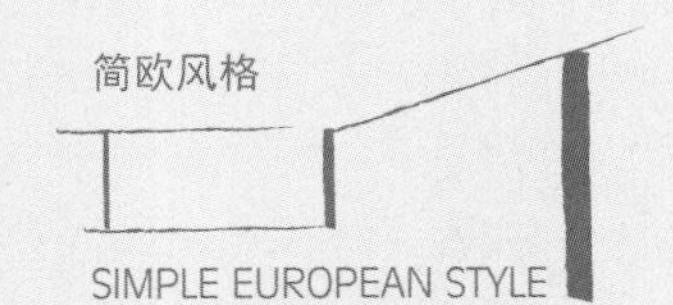

DESIGN CONCEPT 设计理念

“幻想今生只做你的骑士
策马疾驰时间的始末，永不停息
还在古堡幽城中静静的弥望
残云嘲笑着没有尽头的目光，伸向远方
只怕你真的不知道
即使是你的骑士，也会有落泪喘叹的窒息迷茫”。

The story of The Knight's Tale gives us unlimited romantic reverie. Sometimes we dream of living in the stories.

西方中世纪有关于骑士的传说，使得人们充满了浪漫的遐想。现实似乎乏味、单调，总是希望手持长剑，驾于马背，终身为正义而战，胜利归来能陪着美丽的公主居住在古色古香的城堡中，过着贵族一般富足的生活……

We write down this story in a romantic way which sounds like beautiful symphony.

于是，我们的故事开始了。在现代都市中营造这样的氛围，我们将浪漫的向往写于当下，新的碰撞写下交响的乐章。

ROMANTIC METROPOLIS
浪漫大都会

项目名称：沈阳华润二十四城　设计公司：李益中空间设计　硬装设计：李益中、范宜华、董振华　陈设设计：熊灿、欧雪婷

施工图设计：叶增辉、袁晖　项目地点：辽宁沈阳　项目面积：309 平方米

主要材料：银白龙大理石、水纹银大理石、欧亚木纹大理石、黑色橡木木饰面、深色木地板、古铜钢、灰色绒皮、浅灰色布艺硬包等

For this project, we focus on communion between classic and modern; intersection of the castle and city. This art place seems to be the symphony of painting, photographs, music, drama, wine and cigars.

沈阳华润二十四城之浪漫大都会定位于“古典与现代的交融；古堡与都市的交汇”，空间散漫、充满艺术情调，是绘画与摄影，音乐与戏剧，红酒与雪茄的交响。

A large outdoor garden, a bar, a cigar room as well as a media room are all included base on other necessary space in order to symbolize class and elegance of the client's.

在空间功能布置方面，我们在满足基本的生活功能需求的同时，特别增置了户外观景庭院，以及室内的水吧娱乐区、酒窖、雪茄房和影音室，突显了主人的身份和品位。

无障碍坡道
休闲露台
衣帽间
客房
厨房
公卫
西厨
玄关
盥洗室
过道
小孩房
客厅
餐厅
主卧
玄关
林荫小道
书房
中空
水景
观景庭院
庭院玄关

To show the client's elegant yet unassuming personality, we use a plenty of matte black wood finishes and hard fabric with unique texture including a little bit leather materials. The golden metals meadow the space reflect great modern temperament. The classic lines mixing with the modern design technique successfully make the space become a classic and modern metropolis.

空间中运用了大量的哑光黑色木饰面、独特质感的硬包，我们利用局部的皮革衬托，体现出主人及空间的高雅内涵，却不张扬。而空间中点缀的金色亮光金属，亦是一种时尚大都会的体现。古典线条的运用，与其他现代设计手法的交织和亮面的点缀，整体空间体现出一种古典与现代都会的气质。

The shades of whole program are red, black, grey and cobalt blue which bring special grade, matching multiplicate cultural characteristic.

项目的整体色调则采用红、黑、灰及钴蓝色调来丰富空间的品位，再搭配多元素的文化特质。

Dreaming of romantic nostalgia and modern life, we complete this luxurious and fashionable masterpiece.

多元化的思考，将怀古的浪漫情怀与现代生活相互交融，既具有华贵典雅又具时尚现代。

FURNITURE DESIGN
IDEA BOOK 08
INHOME

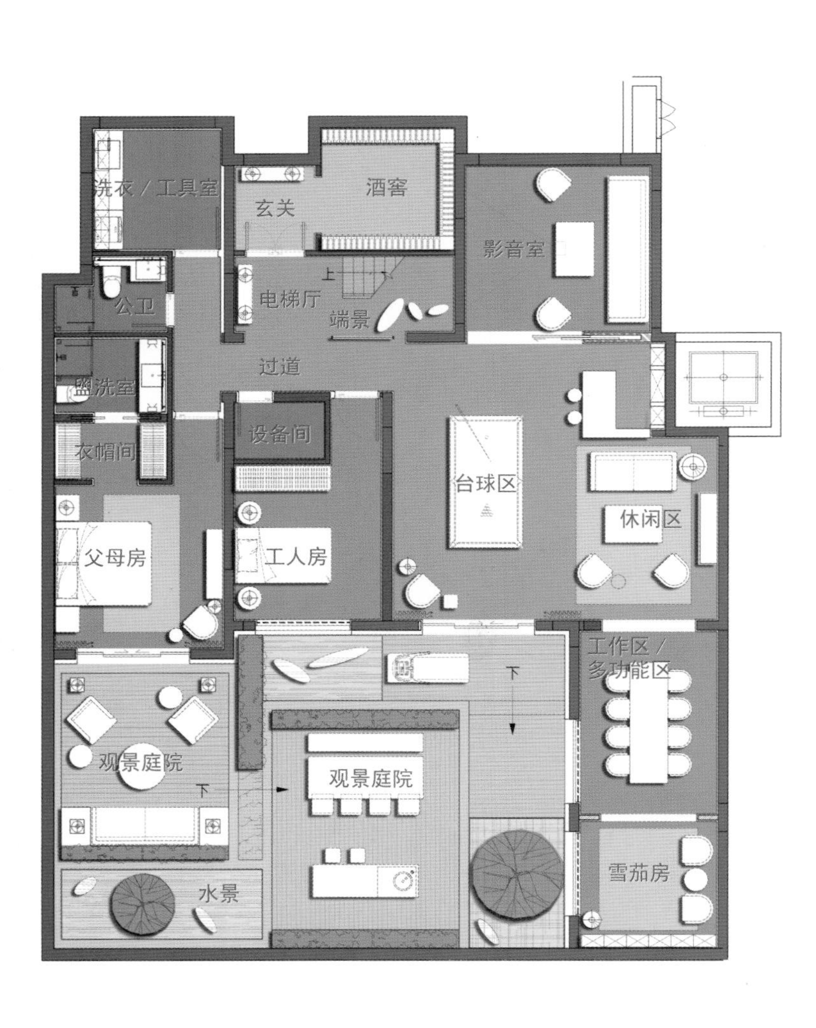
洗衣／工具室
玄关
酒窖
影音室
公卫
电梯厅
上
端景
过道
盥洗室
设备间
衣帽间
台球区
休闲区
父母房
工人房
工作区／
多功能区
下
观景庭院
观景庭院
下
水景
雪茄房

ROCKY

JOHN PAUL GEORGE RINGO
GEORGE RINGO JOHN PAUL
ROCK

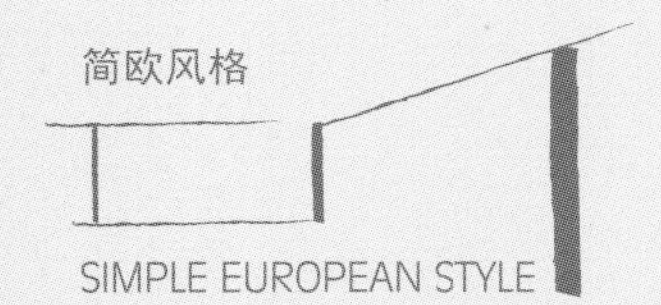

DESIGN CONCEPT 设计理念

The show flat is in simple European style. In order to create elegant temperament and romantic feelings of European style, the designer adopts gray as the main tone, and integrates elegant white and graceful gold in every corner of the space. At first sight the lines in the space are very simple, but after careful savoring, you can feel that the lines make the space flexible and lovely. In the complicated and changeable world, simple and natural living space can make people feel peaceful and cozy. The designer creates an attractive and romantic space by destructing and regrouping the interior space. In the elegant space, every moment can be comfortable.

本套样板间为简欧风格设计作品。为营造欧式优雅的气质和浪漫的情调，设计师以灰色为主色调，将典雅的白色和优雅的金色融合在空间设计的每一个角落。整个空间线条设计初看十分简单，但当你细细品味一番，就能充分感受到线条使空间溢满灵动感。在繁杂多变的世界里，简单、自然的生活空间能让人感受到居家生活的宁静和安逸。设计师借着室内空间的解构和重组，缔造出一个令人心弛神往的浪漫空间。在这精灵般优雅的空间里，每一刻都如沐春风。

ROMANTIC AND ELEGANT URBAN SPIRIT

都市精灵，浪漫优雅

项目名称：昆明海伦国际 301 样板房设计　　设计公司：朗昇建筑空间设计

项目地点：云南昆明　　项目面积：137 平方米　　摄影师：大兵

主要材料：大理石、涂料、木饰面板、瓷砖、墙纸等

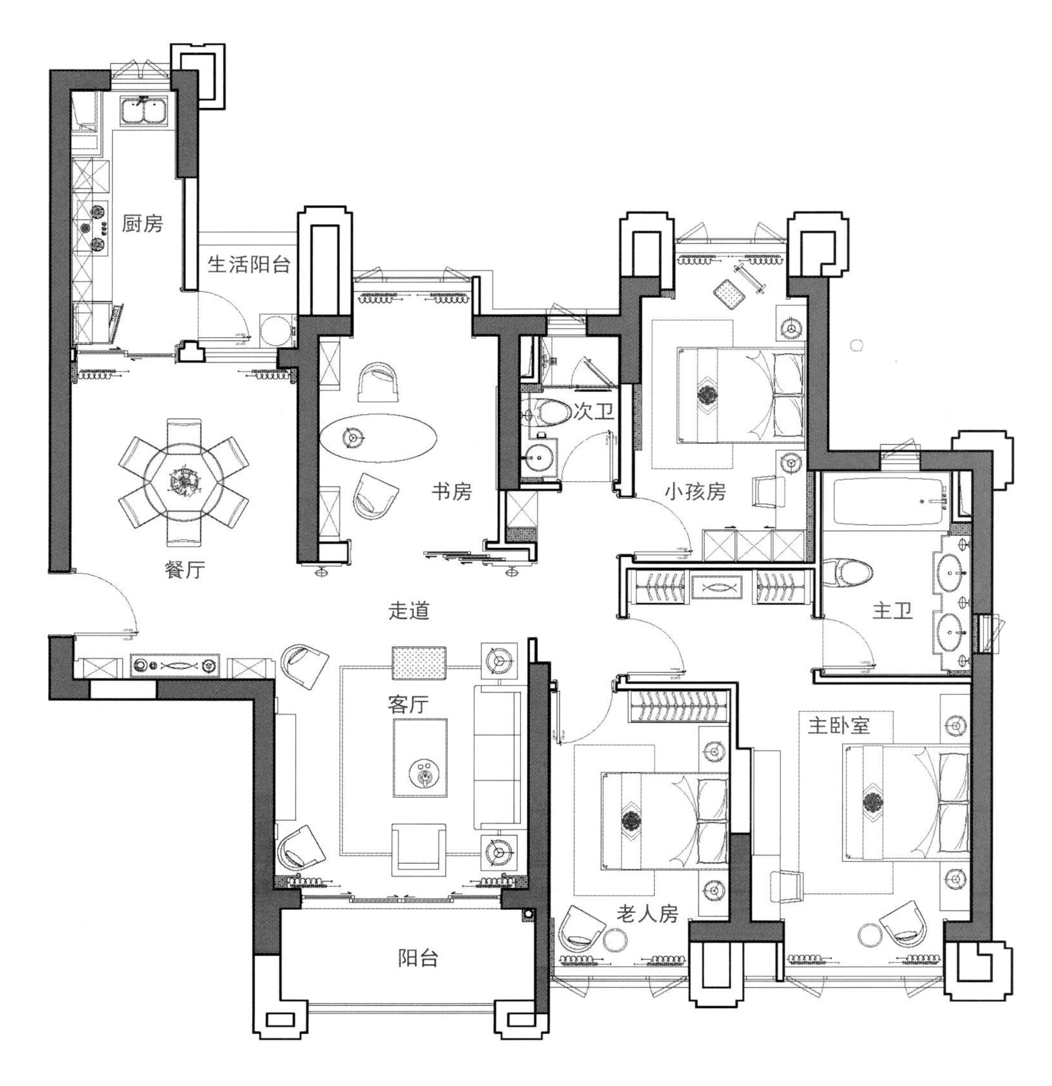
厨房
生活阳台
次卫
小孩房
书房
餐厅
走道
主卫
客厅
主卧室
老人房
阳台

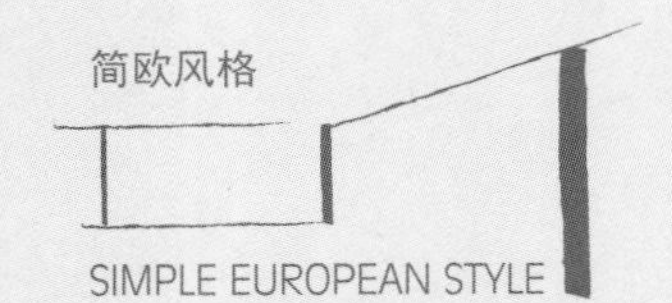

DESIGN CONCEPT 设计理念

The show flat is in European rural style with modern elements and several Oriental elements. The main tone of the whole space is blue and yellow, neutralized by elegant black and white. Romantic European classic blue and yellow silk fabrics collocate with abstract pattern carpet, paintings and lamps, which creates an exquisite, delicate, free and romantic tone, and presents a natural and rural style.

本套样板房设计以欧式田园风格设计为主，与现代元素相结合，并辅以少量东方元素作为点缀，整体空间以蓝、黄色作为色彩基调，使用优雅的黑、白色中和，浪漫的欧式经典蓝、黄色丝质布艺与抽象纹样地毯、挂画、灯具相搭配，营造了精致细腻而又自由浪漫的空间格调，呈现出充满自然气息的田园风情。

THE BLUE IS THE FANCY COLOR OF THE SPACE

那抹蓝，是空间的华彩

项目名称：昆明海伦国际 304 样板房设计　　设计公司：朗昇建筑空间设计

项目地点：云南昆明　　项目面积：120 平方米　　摄影师：大兵

主要材料：大理石、涂料、木饰面板、瓷砖、墙纸等

I can't explain

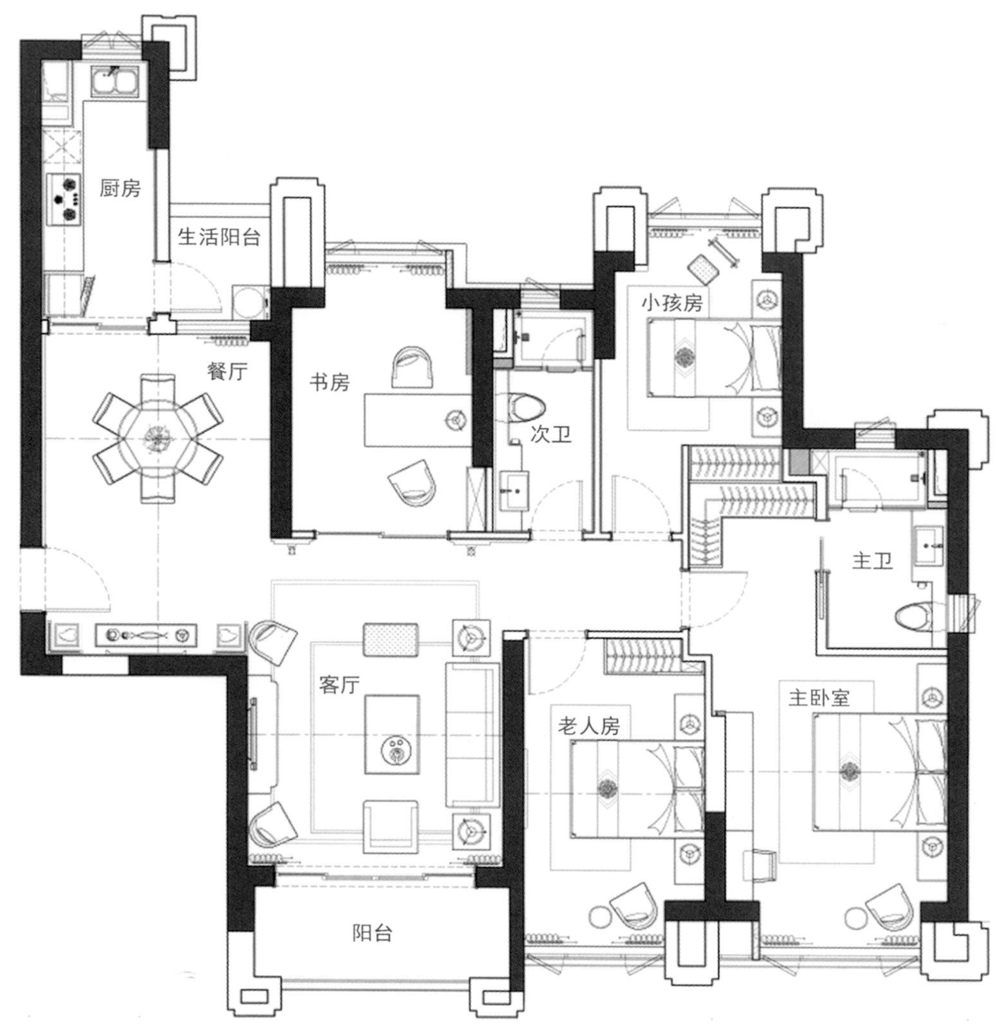
厨房
生活阳台
餐厅
书房
次卫
小孩房
主卫
客厅
老人房
主卧室
阳台

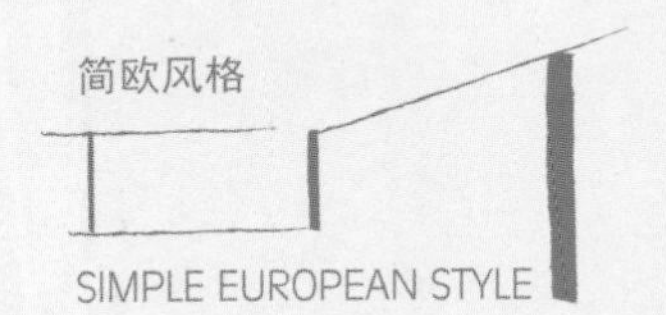

DESIGN CONCEPT 设计理念

He is a software engineer of nine to five in the day time, and a part-time spinning coach in the community gym at night. On the Internet, he is a geek on Zhihu and a writer of a special column. In reality, he is a mother's son, a son's father, David's cousin and her lover.

He says that Peter Pan is the oldest child and Barbie is the smallest adult. The merry-go-round is a curative equestrianism for adult; dream is the ageless beauty care, which only can grow up with curiosity. He says there is a dinosaur in his heart, which is innocent, brilliant and powerful.

白天是朝九晚五的软件攻城狮，晚上是社区健身房的spinning兼职教练。网络上，是知乎上的Geek达人、果壳专栏的写手；现实中，是王妈妈的儿子，娃娃的爸爸，大卫的表弟以及某个她的爱人。

他说：彼得·潘是最老的小孩子；芭比是最小的大人。旋转木马是成年人的治愈马术；梦想是不老的保养品，有好奇心才能继续长大。他说他心中住着一只恐龙怪兽，天真烂漫、充满力量。

THE HOME OF MR. DINOSAUR

恐龙先森的家

项目名称：济南绿地国际城样板间　　设计公司：成象设计　　软装公司：成象软装

项目地点：山东济南

What is the surprise of life? Is it a beam of sunshine from the oven-fresh French bread? Is it the life time memory from the movie? Is it the inspiration from blue mountain coffee?

Life is ordinary fun and continuing passion; life is the turn of seasons and the alternation of orders; life is mutual understanding. Life is love and to beloved.

生活的惊喜是什么？是刚出炉的法式面包附赠的那一束阳光？是看电影时附赠的一辈子的回忆？是喝蓝山咖啡时附赠的一杯灵感？

生活是平凡的乐趣，是续写的热情；生活是季节的更迭，是顺逆的交替；生活是彼此的默契。生活是爱与被爱。

Photos freeze-frame the space and time, and condense memories. Beautiful photo walls decorate the house, and happy memories adorn life. Who is from mountains and lakes, but confined by day and night, kitchen and love? Mr. Dinosaur stands here still always, blessing all departures and welcoming all arrivals. The reindeer is the owner's favorite one, which can collect sundries and make the space flexible. She loves ballet so that he puts ballet girl in the desk. She loves flowers and plants so that he makes flowers bloom in every corner of the house. Life is mutual concerns and understanding. The light is dark; the sleep is shallow. The decorations on the wall are like clouds and weaves, picking up the unimpressive peace of the day after day life. Beautiful things, such as two books, a branch of flowers, a tender deer, always make people warm and happy. The daughter is the father's amour while the father is the daughter's soft spot. Daughter's room is the warmest and pink place in the house. He says with showy facial expression that having a daughter can make time smile.

照片定格时空，凝结回忆。漂亮的照片墙装扮房子，美好的回忆妆点生命。是谁来自山川湖海，却囿于昼夜、厨房与爱？恐龙先生总是安静的站在这里，祝福所有的出发，迎接所有的抵达。既可以收纳杂物，又可以给空间添几许灵动的驯鹿是主人心头之好。她爱芭蕾，他在书桌上摆上芭蕾女孩。她爱花草，他让小碎花开满家的角落。生活，就是相互惦念彼此默契。光线淡了，睡眠浅了，墙面的装饰似云朵如波澜，在日复一日的生活里，拾起那些不起眼的平静。两本书一朵花，温柔小鹿悄然而立，美好事物让人心生温暖喜悦。女儿是爸爸的盔甲，女儿是爸爸的软肋。女儿房是家中最温馨最粉嫩的角落，他说：有女儿，时间会笑，表情炫耀。

HOME

新中式风格

NEO-CHINESE STYLE

DESIGN CONCEPT 设计理念

Home is not only a residence which can keep out wind and rain, but also a harbor which can relieve and console the exhausted heart of the resident after handling heavy things. This project is located in the new district of the city. The owner has a family of four who live in a cultural and artistic environment. The couple who has a successful career and enjoys life, likes playing chess in spare time. Their daughter is good at music. The elder likes painting and tea art. Inspired by the game Go, the designers draw from the vertical and horizontal patterns of the chessboard by using interlaced or interval lines and squares in the space, which forms a diverse expression in the unified sense of order. The space is born for people and also endowed with emotions because of people. This project excepts to frame the whole space with concise design language by breaking the so-called restriction of style, supplemented by Chinese image elements to break the boundary between modern and traditional and make an integration, which endows the space with concise and lively modern sense, keeps links with urban life and conveys the resident's better expectations of pursuing a high lever spiritual life.

CHESS AND FUN

弈 · 趣

项目名称：嘉华广场（二期）住宅项目 3 栋 13 层 03 单元复式样板房

设计公司：中海怡高建设集团股份有限公司、广州市联智造营装饰设计有限公司　设计主创：李雷夫、叶颢坚

项目地点：广东广州　项目面积：211 平方米　摄影师：隐象建筑摄影——欧阳云

主要材料：仿石砖、工艺夹丝玻镜、欧洲灰镜、亚黑拉丝不锈钢、浪琴灰大理石、新土耳其灰大理石、夹画玻璃等

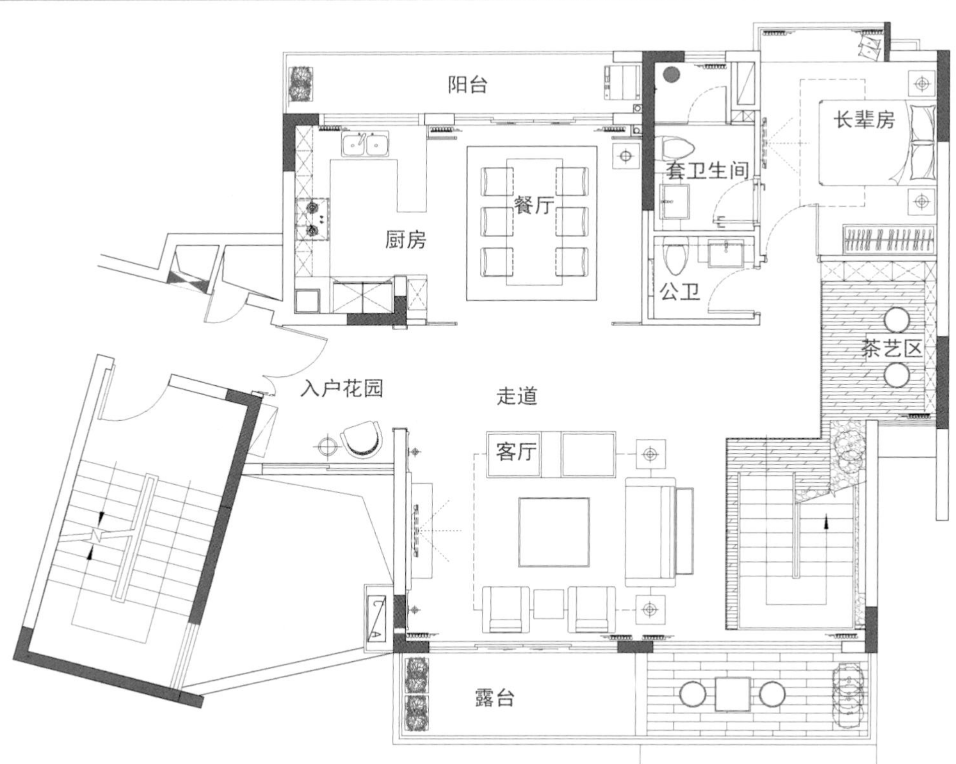

阳台
长辈房
套卫生间
餐厅
厨房
公卫
茶艺区
入户花园
走道
客厅
露台

家，不仅是遮风挡雨的居所，更是能让居者与繁重世事博弈后，疲惫心灵得以舒缓、慰藉的港湾。本案位处城市新区，业主背景设定为富有文艺氛围的一家四口：男女主人事业有成也乐享生活，闲暇时喜欢捉子对弈来上一局；女儿精通曲乐；老人喜爱绘画和茶艺。从围棋获取灵感，借用棋盘上的纵横形式，以线条或方格的形式在空间中穿插运用，或交错或间隔，在统一的秩序感中又形成不拘一格的表现力。空间因人而生，也因人而被赋予情感，本案希望能通过突破所谓风格的限制，以简练的设计语言为空间整体框架，辅以中式意向元素，打破现代与传统的界限，相互交融，既赋予空间简洁明快的现代感，保持与都市生活的连接，也传递出居者对精神生活更高层次追求的美好愿景。

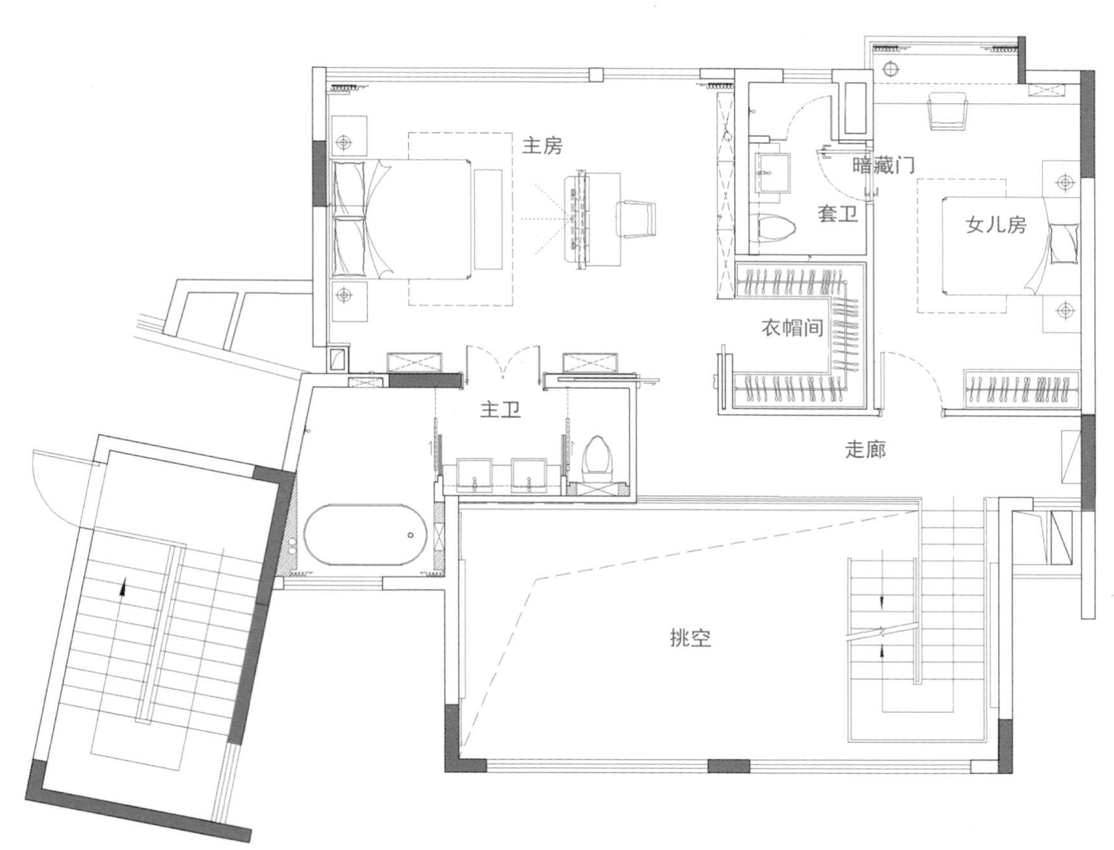
主房
暗藏门
套卫
女儿房
衣帽间
主卫
走廊
挑空

DESIGN CONCEPT 设计理念

Because the design, implementation and sale are on midsummer, so when choosing materials and colors, the designers adopt low-key senior ash to creating a towering, elegant and tranquil artistic conception to bring a touch of cold to the hot summer. In order to continue the simplicity insisted by modern Orient, the concise and skinny lines outline the Oriental spirits. "Senior ash", "mountain" and "cold" are the key words of our design concept. The emotional space changes as seasons change, the scenery in which also changes. Every branch is a blooming soul in nature. The vitality makes the living fun. People here naturally feel the enjoyment brought by the space. Comfortable experience can unconsciously be a beautiful memory. Surprise is far from over. The designers use Chinese methods to partition the space, as the famous architect Ieoh Ming Pei said "let the light design".

SILVER MOUNTAIN SPRING

银色山泉

项目名称：茂名保利海湾城 D 型楼王　室内设计：C&C 壹挚设计　软装设计：C&C 壹挚设计　主创设计：陈嘉君、陈秋安

参与设计：邓丽司、贺岚　项目地点：广东茂名　项目面积：180 平方米

因为项目的设计、实施和开售均正值盛夏，因此我们在选材以及配色都采用了低调的高级灰，希望能营造出一种高山仰止，淡雅宁静的意境，为炎热的夏天送来一抹清凉。为了延续摩登东方所坚持的简约，通过简练骨感的线条勾勒出东方的精气神，而“高级灰”、“山”及“清凉”都是我们设计概念的关键词。这个充满情绪的空间跟随着季节变换，里面的风景也轮番着呈现。每一枝干茎，都是大自然里盛开着的灵魂。几处生机，让居住产生趣味。正在其间的人，自然地感受着空间带来的快乐，惬意的体验不自觉也会成为一段美好的回忆。惊喜远未结束，设计师妙用中式手法对空间进行切割，用善于玩建筑的贝聿铭的话来说就是：“让光线来作设计”。

CINDERELLA
CINDERELLA
CINDERELLA

CINDERELLA
CINDERELLA
CINDERELLA

When people are extending individuality, the tranquility here is a rest symbol to the outside noise, and a rest for freely showing different inner emotions. Regardless of the trend of loving the new and loathing the old, the designers keep their calm personality. The funny combination of Chinese and modern style again bursts out powerful vitality. The relaxing state coincides with the leisurely and carefree mood here. The neat lines conform to the magnificence highlighted by Chinese style. The abstract ink leaves an illusion of traveling through time. The massy texture of materials such as black steel and Acrylic add modern sense, otherwise how can people get away from this ethereal and graceful space? The combination of modern and Chinese style makes the holiday space more interesting and amusing. The changeable pictures and the progressive carpet bring Oriental beautiful and elegant artistic conception of Zen. It is also the partly hidden and partly visible delicacy of seeing big things from small ones, making complicated things simple.

在人人张扬个性的时候，这里的宁静是面对外界喧嚣的休止符号，是给我们心中各自不同的情调自由流露的休憩。不顾潮流的喜新厌旧，设计师保持了自己内心沉稳的性情。中式和现代有趣的结合又一次迸发出强大的生命力，轻松的状态正好与这儿的闲情逸致匹配，线条规整符合传统中式强调的大气。不具象的水墨感总给人以时空穿越的错觉，但是拉丝黑钢和亚克力等材质坚实的触感增加了不少现代感，否则让人如何能够从这空灵逸远的空间抽身？现代和中式的结合让这个度假型的空间显得有趣好玩。虚实幻变的挂画，层层迭进的大地毯带来东方隽逸的禅意，也是以小见大、化繁为简、似有若无的精巧。

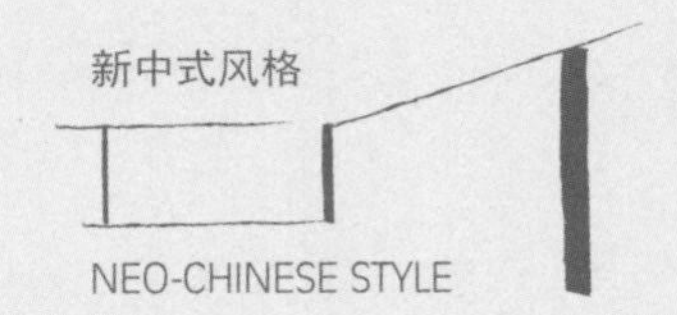

DESIGN CONCEPT 设计理念

Modern style combines with contemporary Oriental fashionable art form, which injects Oriental space spirits into modern space to create a lifestyle featured by contemporary people. It is like that the simple combination of one piece of paper and two points of ink can express the most exquisite feelings of people. The white paper with bold splash-ink and slender lines perfectly combines aesthetics with feelings, which is fashionable, concise and romantic. The promotion of the original architectural structure makes the space balanced. Based on elegant warm colors, different materials intersect, which is comparable and harmonious. The flexible application of modern design methods rearranges some traditional decorative symbols to make them reflect Oriental modern cultural connotations and closer to contemporary people's aesthetics. The whole space stresses the combination of style and art. The furnishings are authentic and elegant, which creates an ideal living space. On the foundation of modern, concise, exquisite and rigorous hard decoration,

INFINITE IMAGINATION, THE BEAUTY OF ARTISTIC CONCEPTION

无限遐想，意境之美

项目名称：意境之美——深业新岸线	设计公司：维塔空间设计	设计师：刘孟
项目地点：广东深圳	项目面积：140 平方米	摄影师：江河室内摄影

the new Zen-like method makes a collision between Oriental "spirit" and modern "form", which concretely presents an aesthetic space with more dimensionality and depth, and custom makes a new residence which is suitable to live and emits Oriental charm for the owner to taste art and enjoy life.

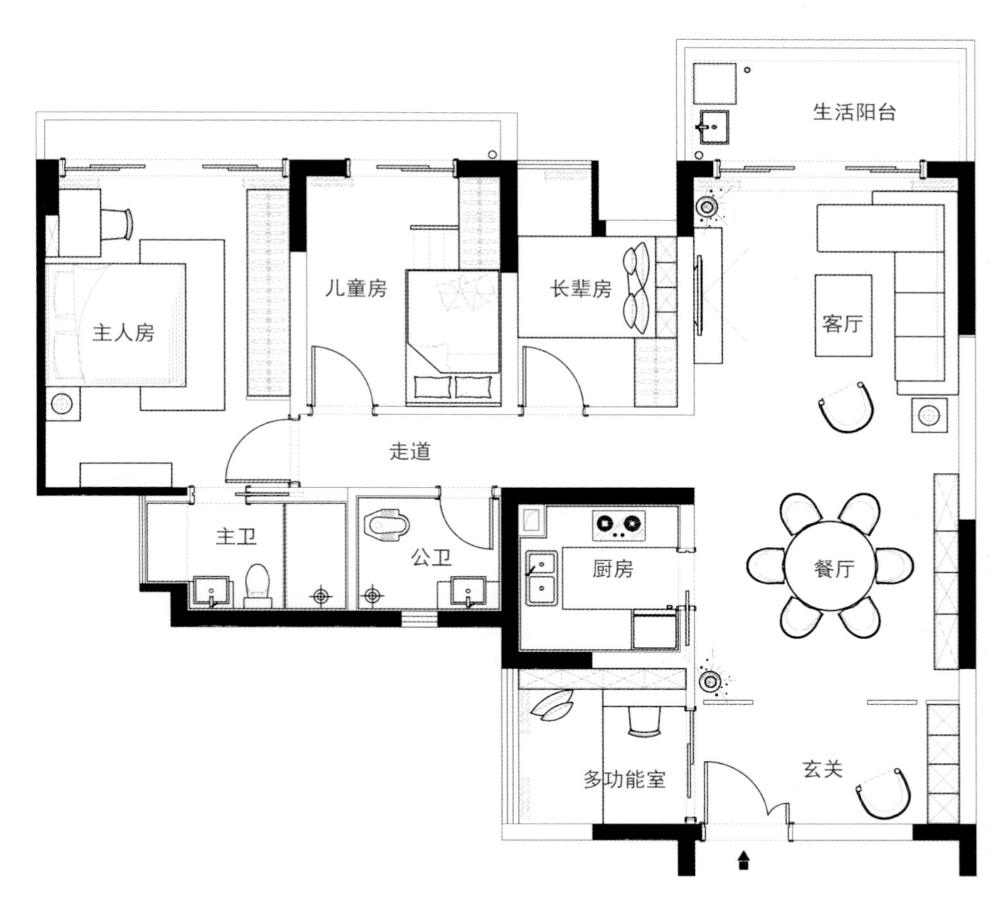
生活阳台
儿童房
长辈房
主人房
客厅
走道
主卫
公卫
厨房
餐厅
多功能室
玄关

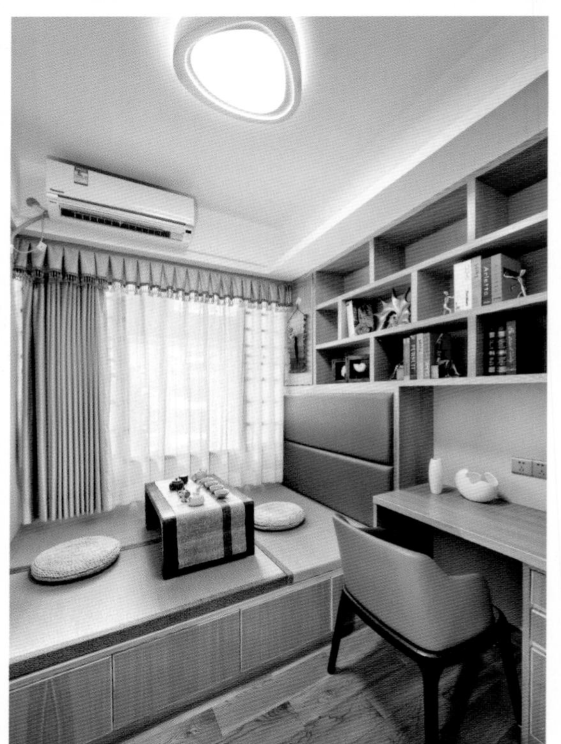

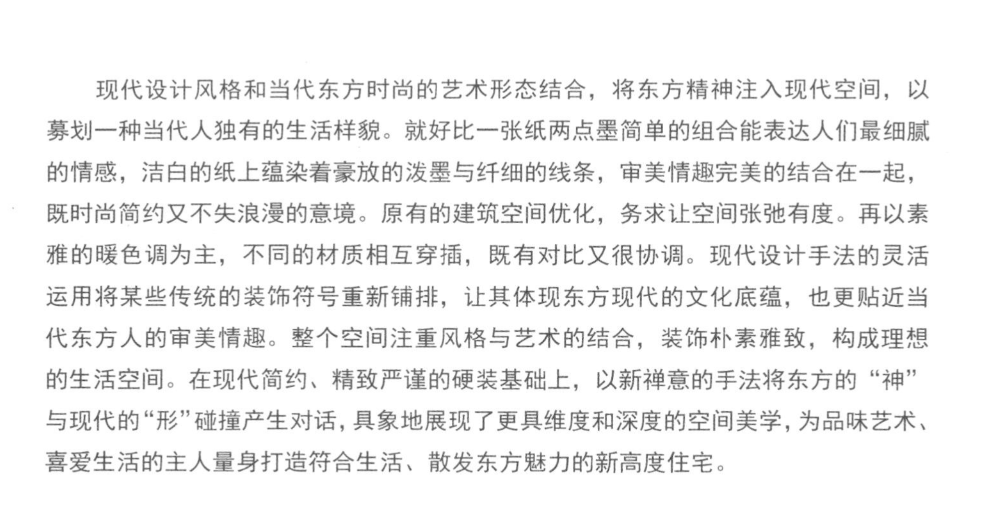

现代设计风格和当代东方时尚的艺术形态结合，将东方精神注入现代空间，以募划一种当代人独有的生活样貌。就好比一张纸两点墨简单的组合能表达人们最细腻的情感，洁白的纸上蕴染着豪放的泼墨与纤细的线条，审美情趣完美的结合在一起，既时尚简约又不失浪漫的意境。原有的建筑空间优化，务求让空间张弛有度。再以素雅的暖色调为主，不同的材质相互穿插，既有对比又很协调。现代设计手法的灵活运用将某些传统的装饰符号重新铺排，让其体现东方现代的文化底蕴，也更贴近当代东方人的审美情趣。整个空间注重风格与艺术的结合，装饰朴素雅致，构成理想的生活空间。在现代简约、精致严谨的硬装基础上，以新禅意的手法将东方的“神”与现代的“形”碰撞产生对话，具象地展现了更具维度和深度的空间美学，为品味艺术、喜爱生活的主人量身打造符合生活、散发东方魅力的新高度住宅。

DESIGN CONCEPT 设计理念

Zen is a realm of life and also an experience which can equip you for the whole life.

禅，是一种生活境界；禅，又是一种受用，一种体验。唯有行者，唯证者得。

Zen is a mundane meditation, which emphasizes you to master inner peace. "Void" (in Buddhism) reaches the highest realm. Our hearts should not be attached to anything. Since the new realm appears, mind and matter have combined emotion and object into one.

禅本是静虑、止观的意思，强调心灵的参悟，它的最高境界乃是"空"，让人追求心无挂碍的灵魂悟诗与禅结合。自有新境界出现，即心与物交融而使美的情感与物象合一。

For this project, there are four bedrooms and two living rooms with 154-square-meter floor area. It's elegant, transparent and the size is quite square which can be fully used.

该项目户型建筑面积 154 平方米，四室两厅居室，户型方正，使用率高，大气通透，开阔灵动。

ZEN ART
禅风艺境

项目名称：沈阳华润二十四城　设计公司：李益中空间设计　硬装设计：李益中、范宜华、关观泉

陈设设计：熊灿、欧雪婷、李晴　施工图设计：叶增辉、胡鹏　项目地点：辽宁沈阳　项目面积：148 平方米

主要材料：橡木地板、翅木饰面板、黑色拉丝不锈钢、白色人造石、白金沙大理石、墙纸、布艺硬包、园林木地板等

The semi-open bar with the wooden partition brings space a sense of order and sets off the landscape arts, which plays with the effect of finishing touch. The living room with perfect symmetry has exquisite and simple lines, covering carpets with dried mountains and rivers on the floor. We express idea with the shape, and express emotion with the idea.

吧台半开放隔断强烈的秩序感，映衬画龙点睛的造景艺术。客厅完美的对称，细腻简洁的线条，形似枯山水的地毯，以形写意，以意传情。

The dining room faces the balcony, then client can see the beautiful view through windows anytime.

餐厅正对着外景的大露台，用餐时可观赏窗外美景，心情愉悦舒畅。

Home sweet home. We hope to create a quiet and cozy place where is not only relaxing but returns to the natural instinct of life.

家，是一个温馨的地方。在这繁华的都市中，我们希望它具有宁静的氛围，来使得这个家变得安详惬意，让人全身心得到伸展，从而回归本性。

“Avalokita, the Holy Lord and Bodhisattva, was moving in the deep course of the Wisdom which has gone beyond. He looked down from on high and beheld but five heaps, then he saw that in their own-being they were empty”, mentioned in The Heart Sutra. You can release from worries and pain when you are void and gain great wisdom from peace and quiet.

“行深般若波罗蜜多时，照见五蕴皆空”，也就是无我的时候，一切烦恼与痛苦得到解脱，继而在清净中获得大智慧。

Someone says the meaning of Zen style is to create the supreme wisdom and to prove the wisdom of everything.

“东方意境的禅风，禅的意义就是在定中产生无上的智慧，以无上的智慧来印证，证明一切事物的真如实相的智慧，这叫作禅。”

Space not only contrasts environment but also is client’s understanding and attitude towards humanities.

空间既是作为环境的烘托，也是业主对人文的理解和态度。

Oriental Zen style design also satisfies art appreciation, which not only pares down to the essence but simple and generous. The space with clean and crisp lines is strong yet pliant but has a combination of strength.

东方禅风也非常符合艺术审美，它去繁从简，界面简洁大方。空间线条干净利落，柔中带刚，刚中带柔。

客房
茶室
公卫
厨房
吧台
玄关
主卫
衣帽间
小孩房
客厅
餐厅
主卧室
阳台

The regular and balanced space layout brings balance and harmony. Warm colors and soft lights highlight a sense of refinement.

方正对称的空间布局，给整体带来更多的平衡及协调感。色彩温暖，灯光柔和，简单中彰显精致。

DESIGN CONCEPT 设计理念

Walking into the Town of Brocade at the first time, there is a tranquil, fresh and elegant feeling. Pink walls and black tiles, front and back yards, and the surrounding water channels form the beautiful environment with "small bridges, flowing streams and homes". The exterior walls are closed, natural and rustic, whose lines are well-arranged, covert, elegant and simple. The façade with Anhui-style architectural features is based on the principle of "learning from nature and following the form" and the owner's requests to life. The designers position the interior design style of this project as Anhui-style Neo-Chinese style without hesitation. According to the owner's aesthetic needs and the designers' understandings of the architecture itself, the designers cerate a Neo-Chinese style different from the traditional one. They keep the tranquility and depth of Jiangnan ink painting, and boldly use Chinese furniture with gilded gold. The gold lines of curtains and gold embellishments of furnishings present exquisite, noble, heavy and elegant feelings. The unity of these two creates a new light luxurious Jiangnan style.

THE TOWN OF BROCADE

锦镇年华

设计公司：成都优家慢享装饰设计有限公司	主创设计：孙静	参与设计：韩敏
项目地点：四川成都	项目面积：400 平方米	摄影师：季光
主要材料：石材、实木墙板、定制墙纸、玻化砖、实木地板等		

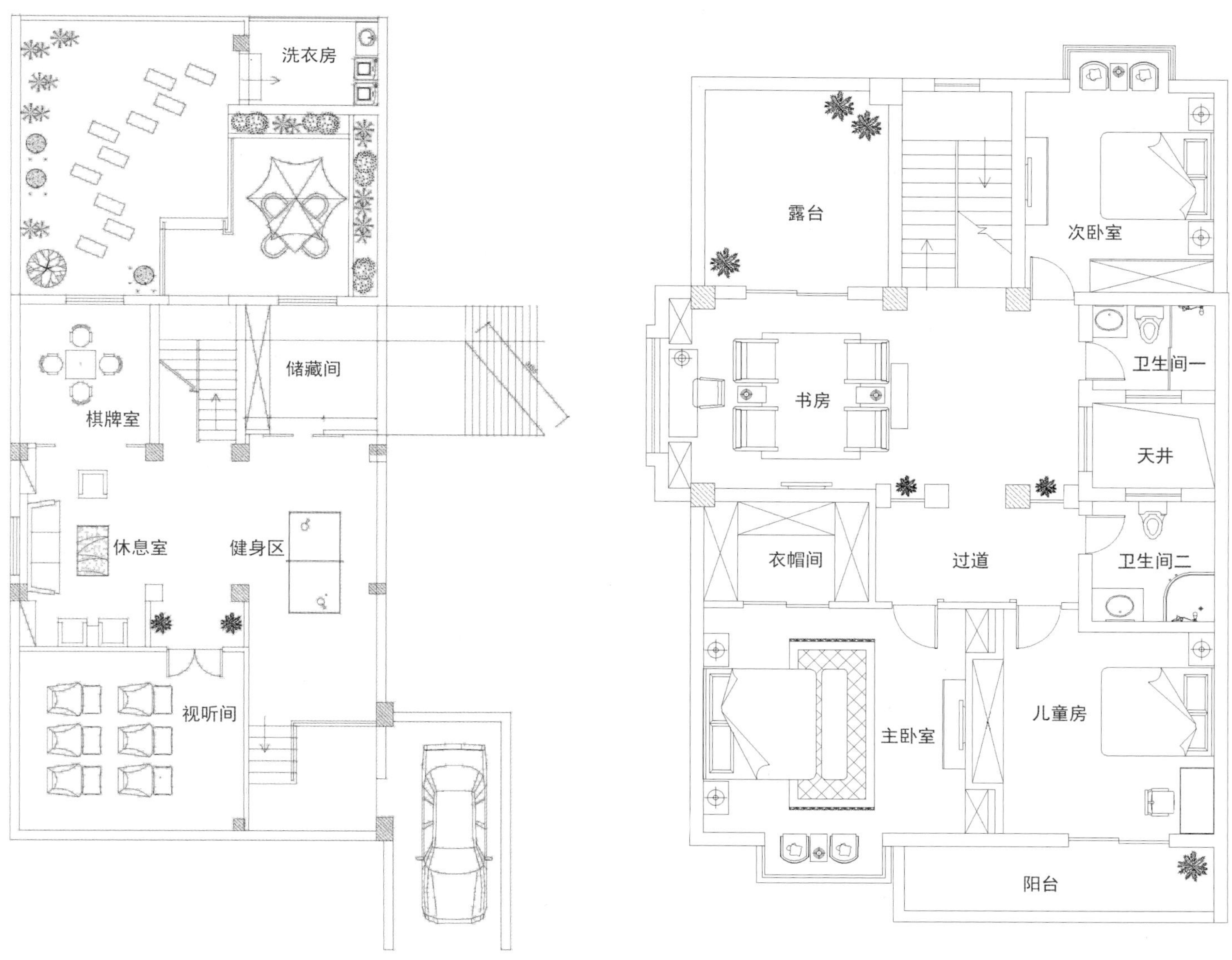
洗衣房
储藏间
棋牌室
休息室
健身区
视听间
露台
次卧室
书房
卫生间一
天井
衣帽间
过道
卫生间二
主卧室
儿童房
阳台

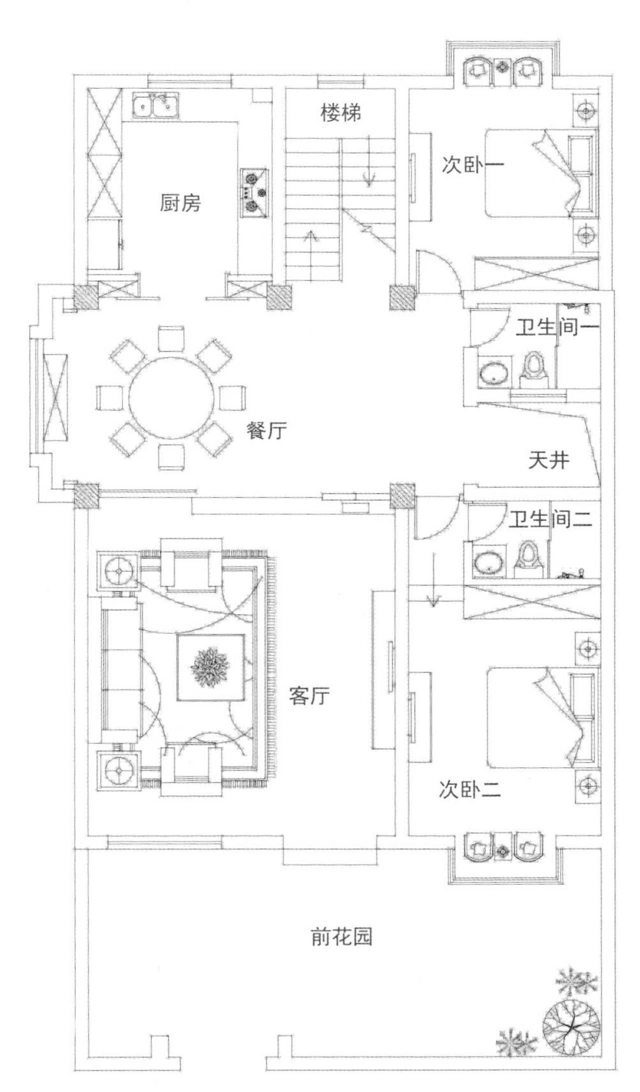
楼梯
次卧一
厨房
卫生间一
餐厅
天井
卫生间二
客厅
次卧二
前花园

露台
露台
连廊
天井
起居室
主卧室门厅
卫生间
主卧室
衣帽间

初次走入锦镇就有一种幽远、清新典雅的感觉：粉墙黛瓦，前庭后院，环绕的水渠构成了“小桥、流水、人家”的优美画面；外观高墙封闭，自然古朴，墙线错落有致，隐僻典雅，不矫饰不造作。这种有点徽派建筑特色的外观及本着“师法自然，顺乎形式”的原则，和客户对生活的要求，我们毫不犹豫地将室内的装修风格定位为：徽派新中式风格。本案以客户的审美需求和我们对建筑本身的理解，打造与传统意义不一样的新中式风格。我们将现代元素和传统文化元素结合在一起，保留了江南水墨清静幽远的特色；也大胆地用描金的中式家具、窗帘的金色勾边、饰品的金色点缀，体现出精致贵气的浑厚高雅之感。将两者高度统一，打造出了全新的轻奢江南风格。

DESIGN CONCEPT 设计理念

This project is based on the study-abroad background of home culture of overseas Chinese in Jiangmen. People here are familiar with cordial feelings of study-abroad culture. The simple neo-classical and western style cannot meet the pursuit of modern people's distinct life feelings. People returning from abroad can possess the Eastern and Western culture. Their bloods flow dense essences of culture of Lingnan and fuse nutrients of modern Western culture. Therefore, the designers hope that the style of this space is exquisite and amorous, and hope that the infinite space can make the combination of Oriental feelings with Western aesthetics more free and open. In addition to chasing the fundament deeply buried in spirit and keeping the complicit and elegant design essences of traditional Oriental style, it strives to present a modern and cozy space, to make the environment and spirit reach the spiritual and peaceful Oriental aesthetic realm, and to interpret culture and spirits of the space by "significance beyond the image, image beyond the scenery" and "delicacy beyond the flavor, purpose beyond the taste". The identity of emotion, culture and temperament becomes increasingly important in the space.

NATURALLY ENCOUNTER BETWEEN WEST AND EAST

随遇西东

项目名称：江门保利中悦花园楼王	室内设计：C&C 壹挚设计	软装设计：C&C 壹挚设计
设计师：陈嘉君、邓丽司、陈稚聪、贺岚	项目地点：广东江门	项目面积：140 平方米

PLAN ATLAS APARTMENT BUILDINGS

本案扎根于江门侨乡文化的留洋生活背景，这里的人对留洋文化有熟悉的亲切感。单纯的新古典主义或者西方风格都不能满足现代人对于别样生活情趣的追求。留洋归来的人们同时拥有东、西方文化浸染，他们的血液里面既流淌着深厚的岭南文化精髓，又融入现代西方的养分。因此，设计师希望这个空间的风格是精致而风情的，希望通过这个有限的空间把东方情怀和西方审美情趣结合得更加自由、更加开放。追逐深埋灵魂的根本，保留传统东方风格含蓄优雅的设计精髓之外，力求呈现现代、隽逸的空间，使环境和心灵都达到灵与静的东方唯美境界，以“象外之意，景外之象”，“韵外之致，味外之旨”诠释空间的文化精神。情绪、文化、气质的认同感在空间里越发重要。

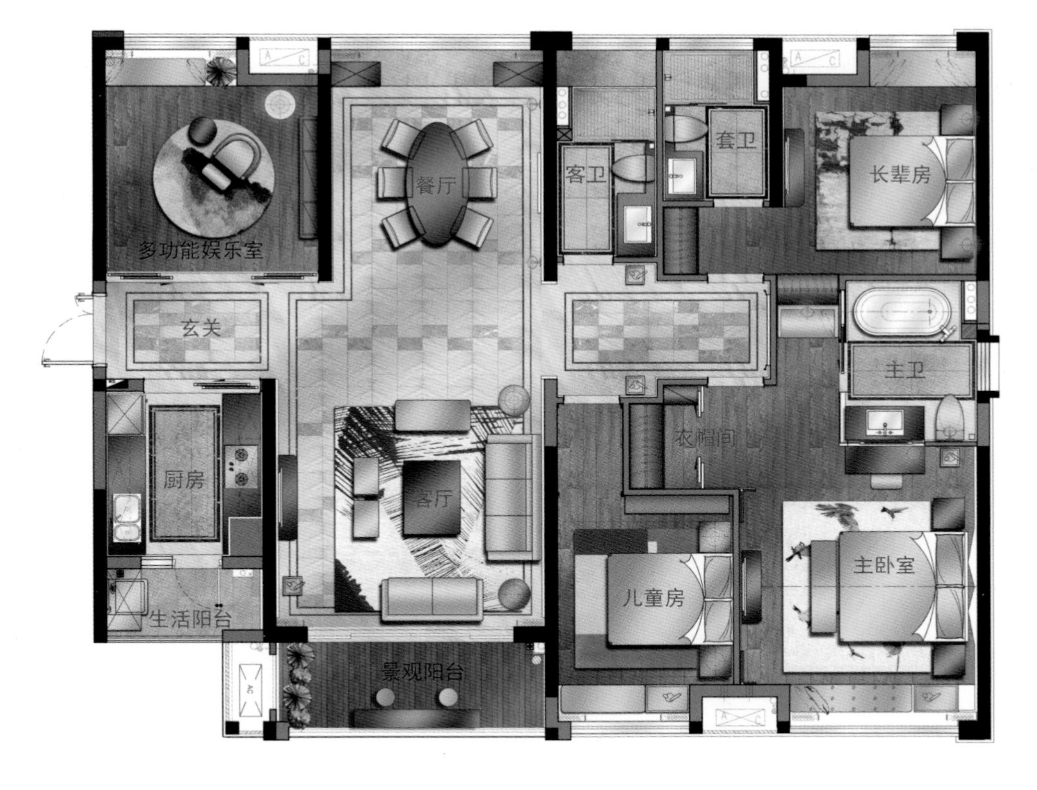

The traditional Lingnan culture is sensitive, delicate and exquisite, and pursues refined elegance as “the colors when the rain passes, the sun shines and the cloud appears”. The lake and sky blue and the jade and glass green fit in with the complicit, elegant and calm demands of Oriental literati, which makes the space flexible through abundant Oriental cultural connotations without wasting one ink. The ebony and ceramic often give people illusions of a long time ago, which makes the cold jades and marbles gentle and elegant. But now the present of a space requires much more so that modern urban breath brought by Chinese and Western culture cannot wait to swarm into. Sometimes it is sedate and magnificent, while sometimes it is low-key and dynamic. The subtle changes between different materials flow in the exotic atmosphere.

传统的岭南文化敏感细腻且精致，追求“雨过天青云破处，这般颜色做将来”这样灵秀脱俗的素雅。湛碧如湖的天色，浅草如玉脂的初青切合东方文人对于含蓄、优雅而沉静的诉求，将丰厚的东方文化底蕴通过空间的变化悄然带动起来，不费一笔一墨。黑檀和陶瓷经常给人回到很久以前的错觉，使得冰冷的玉器和大理石变得温柔儒雅起来。但空间的呈现却不满足于此，华洋带来的现代都会气息迫不及待进涌而入。时而沉稳大气，时而低调跃动，不同材质间微妙的变化流淌在稍具风情的气氛里。

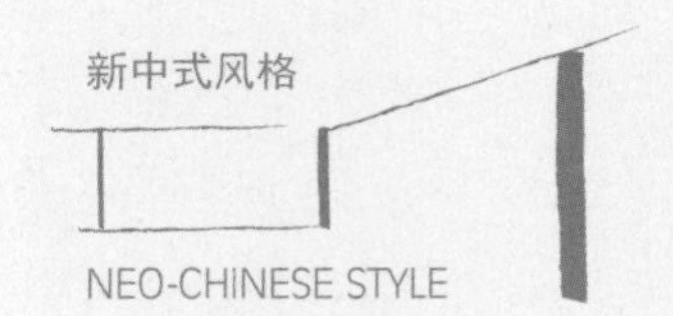

DESIGN CONCEPT 设计理念

Oriental elements in designs tend to be noble and elegant. The elegant light yellow tone with a Zen-like green brings the inner heart back to the quiet and refined world. The heavy Zen-like mood, a clean mind, high mountains, flowing waters and a cup of tea are used throughout the space. The use of natural marbles in the entrance hall is like an abstract landscape painting in the space, which is dynamic and free. The elegant decorative frames with plain lines seem to be casual. The natural environment makes people feel in the quiet garden where the heart can be peaceful.

东方元素在设计中，往往显得高贵而典雅，在优雅的浅黄色调中，加入禅禅的绿意，以回归于内心的宁静清雅的世界。禅意浓浓、心境清清，高山流水、一抹清茶，以“山水、水墨”来贯穿整个空间。入口玄关处天然的大理石运用，就像点缀空间的一幅山水抽象画。灵动而自在，搭配质朴线条优美的装饰架，一切都是那么不经意。天然去雕琢，仿佛置身于桃园幽地，一颗心也被浸润的寂静。

CLOUD, WATER AND A ZEN-LIKE MIND

云水禅心

项目名称：保利江门	设计公司：广州道胜设计有限公司	主持设计师：何永明	参与设计师：道胜设计团队
项目地点：广东江门	项目面积：194 平方米	摄影师：彭宇宪	

主要材料：维纳斯米黄大理石、西班牙玉大理石、山水纹云石大理石、花岗石、红洞石大理石、蒙古黑大理石、山水纹云大理石、深古铜拉丝不锈钢、墙纸、木饰面等

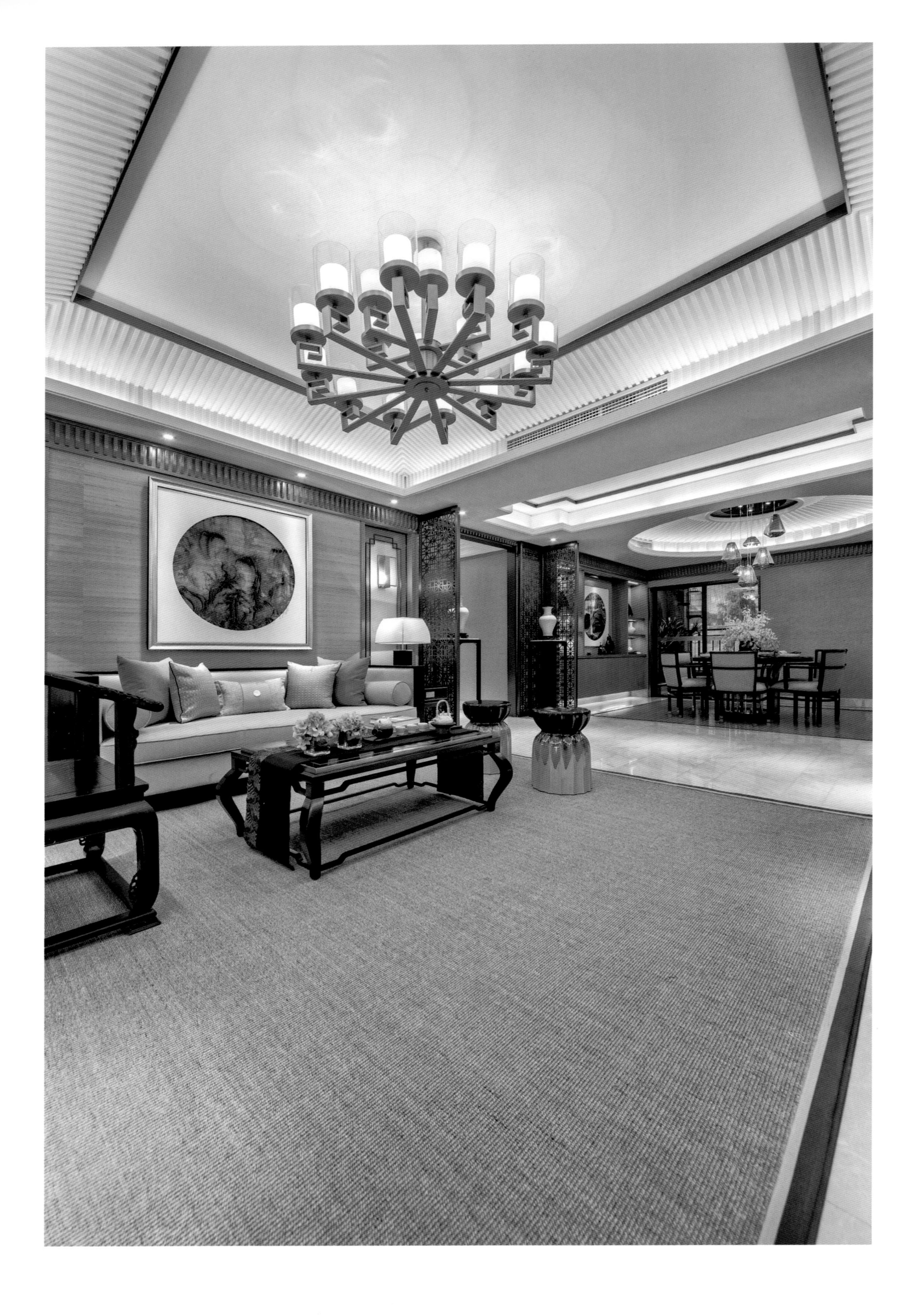

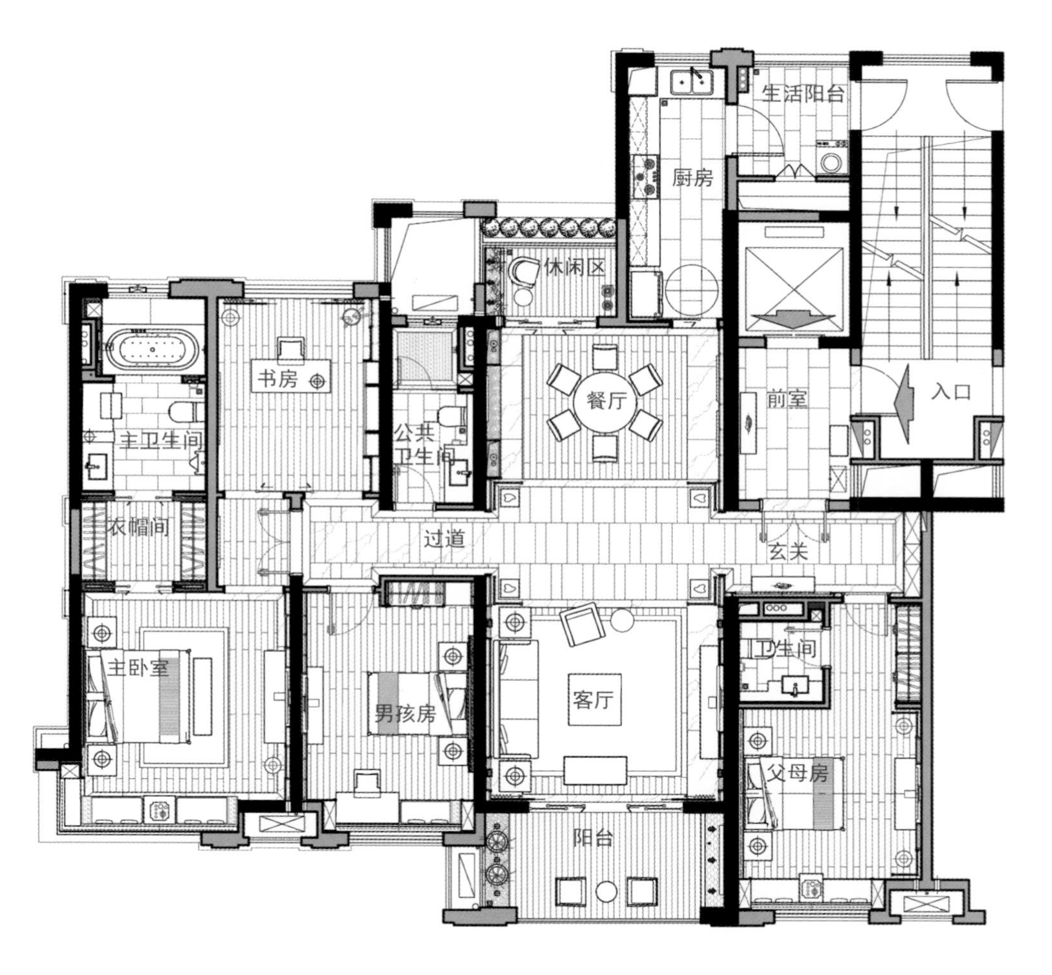

The half open screen in the living room creates transparent and ventilated space and switches the rhythm of the space. Luxurious lamps and exquisite furniture stress elegant flavors. Classical Oriental furnishings, such as flower art, pine tree, four arts, manifest fresh energies, are penetrated into every corner of the space, and are applied into extreme. The master bedroom continues the colors in living room, which emits Chinese elegant temperament. It is not only the superficial art form, but also sensory stimuli which stimulates the spiritual level to resonate. The light penetrates through the gauze in the parents' room without redundant decorations. The tea table and bed are in suitable sizes; the appliances are in diverse patterns; the quiet dark red is flamboyant and implicit. All deduces aesthetics and delicacy, nature and tranquility, and elegance and refinement incisively and vividly. Every corner of the space presents the resident's cultural, noble, free and emotional feelings.

客厅半开放式的屏风隔断营造呼吸通透的量体区间，拨动空间气氛的韵律。豪华的灯具、精致的家具，在设计上强调高雅的韵味。花艺、松树、琴棋书画的古典东方陈设铺设出鲜活的能量，渗透空间每个角落，将中国元素运用到极致。细观主卧，将客厅色彩延续其中，散发出中式儒雅的气质。其不仅仅是流于表面的艺术形式，感官也激发了精神层面，进而产生共鸣。一席亮光透过薄纱的父母房，无需过多饰品矫揉造作的修饰。几榻有度、器具有式，沉寂的暗红色既张扬又含蓄。把唯美与精致、自然与恬静、优雅和儒雅演绎得淋漓尽致，指尖触及的每个角落都能感受到居者的文化感、贵气感、自在感与情调感。

DESIGN CONCEPT 设计理念

A branch of tilted plum blossom, leaning upon a balustrade several times, gray tiles and white walls, a misty rain and a rouge can be exceedingly fascinating and charming.

A branch of tilted plum blossom, leaning upon a balustrade several times, gray tiles and white walls, a misty rain and a rouge can be exceedingly fascinating and charming.

Nature creates things as promised. The hill top is as black as ink; the plain white cloud lingers among the mountains. After a while, researching the free and distinct ink painting and enjoying the luxurious world, you'll find everything is in the variation of shade.

One knows and observes all but stays obscure with concise heart. There is flaw when making ink as if a joke from nature to life. Mountains can never be hidden, and ink painting has elegant and comfortable joy. Landscape is not in the distance but here in the home.

ELEGANT AND COMFORTABLE JOY FROM INK PAINTING

水墨清欢

项目名称：绿地・泉景嘉园复式样板间　　设计公司：成象设计　　软装公司：成象软装

项目地点：山东济南

一支斜梅，几回凭栏，青瓦白墙，一杯烟雨，只添一抹胭脂便可风情万种。

承托载举，布席造境，将一壶置于天地之间，啜一口清香，亦或一抹苦涩，所有凡事便都被茶香消散。

自然造物，总是不负所望，如墨般的玄青是山顶，云雾绕在山间就是素白，半盏时光，研水墨自在，笔蘸性情，品天地山川，世间繁华，却原来都在浓淡之间。

知黑守白，心素如简，砚台调墨偶有瑕疵，也似自然对生活的小小玩笑。青山遮不住，水墨有清欢，山水从来不在远方，此心归处是吾乡。

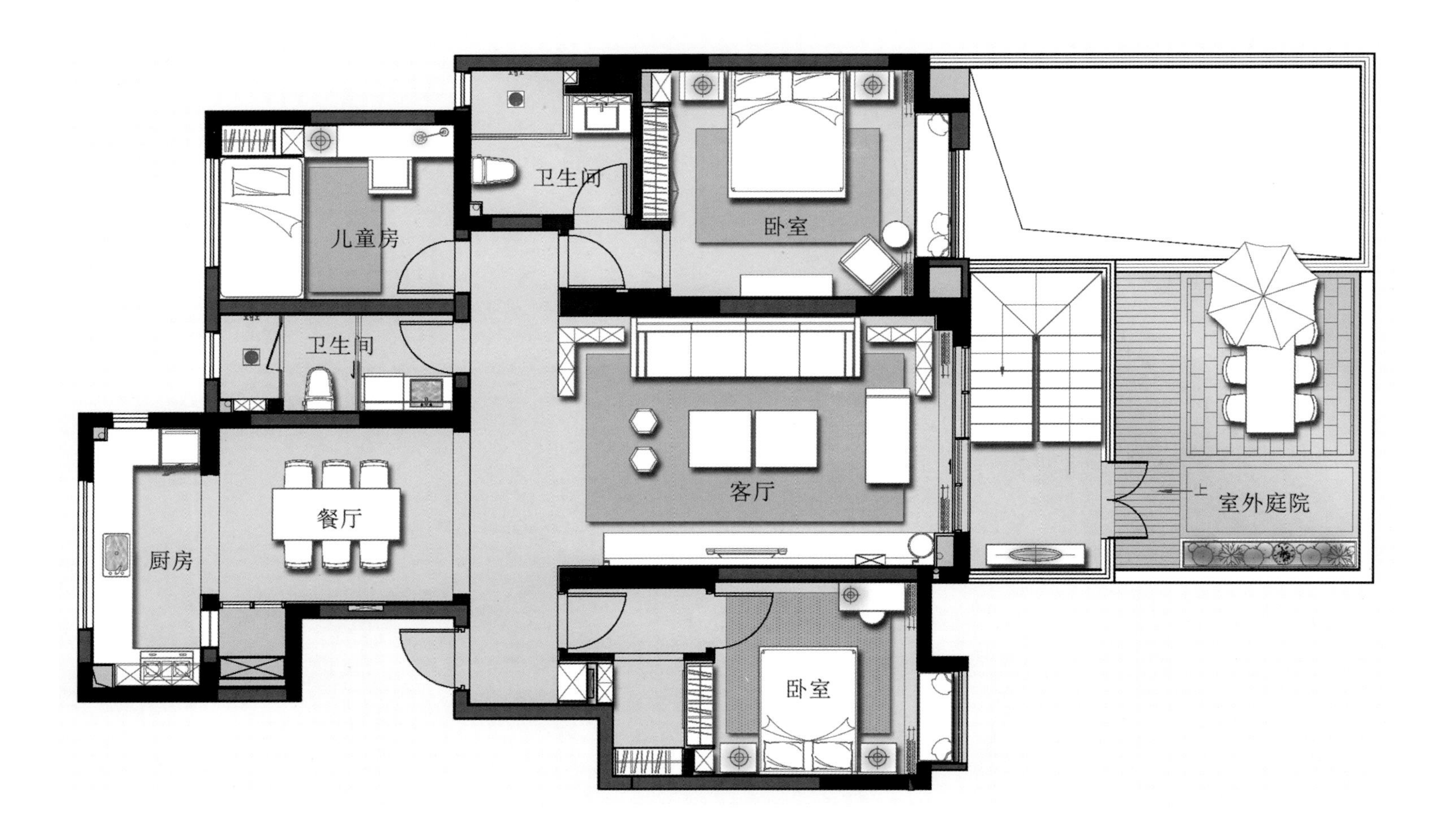

Under the light and shadow, the ink carpet collocates with drum tools as if the clean spring flows in the river. The misty rain is like a painting, which is the owner's favorite Anhui-style. The objects use wood as frames and the walls are decorated with trees. The corridor has dark tiles, white walls and streets as if the ink painting that the owner cannot live without. A clean window, a neat tea table, a warm sun, a gentle wind and a simple diet can be happy. The wind blows through the trees, which makes the house fragrant. Living in the flowers, you can enjoy natural and lovely environment. The tranquil atmosphere and simple and comfortable fabrics create soft artistic state, which is plain and elegant. Graffito, skateboard and riding, the youth should be unruly. Quietly and powerful, this is the warmth of wind. Songhua wine, tea eggs, a cup of tea and a brunch of sunshine, every breath has the fragrance of tea. The night falls, and the mountains are obscure. All is in the ink painting in the house.

承一席光影，水墨感的地垫搭配鼓凳，似清泉石上流。烟雨施如画，不消水墨，不虚浮华，是主人最爱的徽派。物引木为架，墙将树作衣。走廊的布置似青瓦白墙、水墨街巷，这是主人心中离不开的水墨情。窗明几净，暖日和风，只粗茶淡饭，尽有余欢。风吹树花满室香，日居繁花中，自然可爱。质朴与清雅惺惺相惜，宁静的氛围与简单舒适的布艺塑造柔和的意境。涂鸦、滑板、骑行，年轻就要不羁。不动声色却又有让人安心的力量，这是风物的温暖。松花酿酒，煮水煎茶，伴随着杯中的绿意和细碎阳光，每一次呼吸都萦绕着茶香。暮色阴阴，远山淡淡，于室中，皆在一纸水墨间。

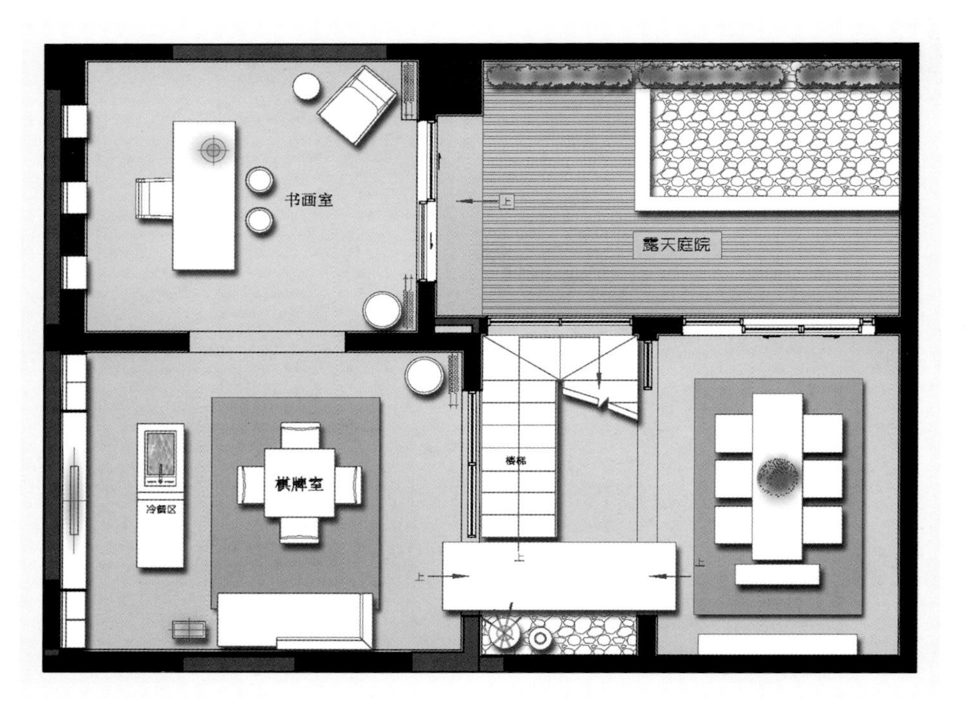
书画室
露天庭院
上
楼梯
棋牌室
冷餐区
上
上
上

Childhood

现代风格

MODERN STYLE

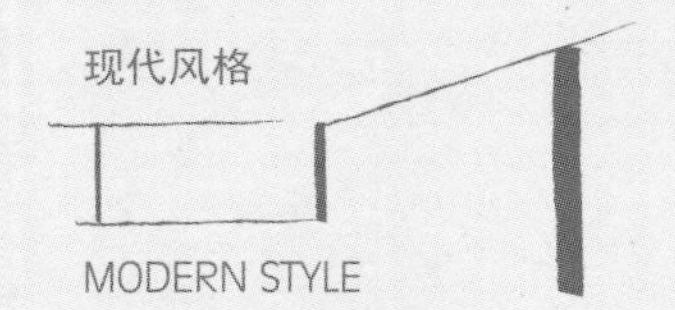

DESIGN CONCEPT 设计理念

Positioned as peninsula style, this project captures elegance and chic from traditional Hong Kong and Shanghai Peninsula Hotel and absorbs essences from the local history and culture to make the owner experience the most authentic French culture. The elevator hall uses elegant cream pinta marbles, collocated with black mirror stainless steel, which forms a sharp color contrast. The rectangular-ambulatory-plane on the ground interprets Art Deco style. The ceiling of the living room uses repeated squares. Meticulous levels and metrical arc wavy lines create rich visual effects. The dark cabinet collocates with bronze stainless steels in the display area, which sets off the colorful and diverse accessories and presents a gorgeous and splendid visual space. The main tone of the family hall is creamy white, collocated with Western paintings, which manifests the owner's noble temperament. The background of the master bedroom abandons complicated textures and furnishings, uses simple upholstery lines, and creates concise and elegant atmosphere by symmetric methods. The bathroom combines ebony veneers with Yves saint Laurent marbles. The walls are largely covered with famous and precious white marbles, which manifests the French peninsula style.

PENINSULA AMOROUS FEELINGS

半岛风情

项目名称：新力洲悦示范单位贰	设计公司：奥迅室内设计有限公司	设计师：罗海峰
项目地点：江西南昌	项目面积：600 平方米	摄影师：FOCUS 摄影工作室
主要材料：黑檀木饰面、白色手扫漆木饰面、黑色钢琴漆木饰面、圣罗兰石、鱼肚白石、白沙米黄石等		

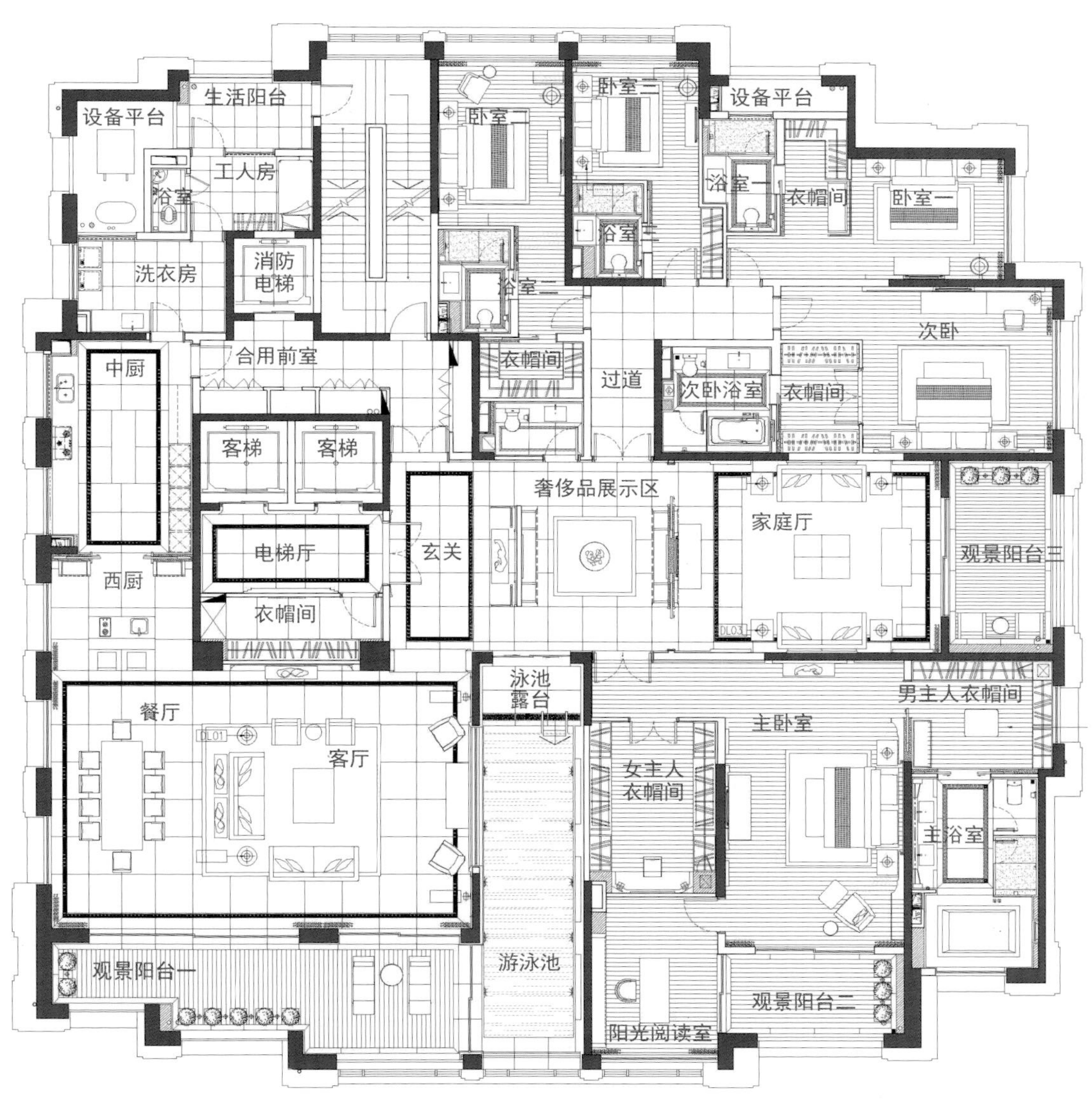
生活阳台
设备平台
工人房
浴室
洗衣房
消防
电梯
卧室二
卧室三
设备平台
浴室一
衣帽间
卧室一
浴室三
浴室二
合用前室
中厨
衣帽间
过道
次卧浴室
衣帽间
次卧
客梯
客梯
奢侈品展示区
家庭厅
观景阳台三
电梯厅
玄关
西厨
衣帽间
泳池
露台
男主人衣帽间
餐厅
主卧室
客厅
女主人
衣帽间
主浴室
观景阳台一
游泳池
观景阳台二
阳光阅读室

本案采用半岛风格，在撷取传统香港和上海半岛酒店的优雅及别致之外，吸收当地历史文化精髓，使户主体现最真实的法式文化。电梯厅运用淡雅的白沙米黄石配搭黑色镜面不锈钢，颜色对比鲜明，地面回字形的地花更能体现 ART DECO 风格。客厅天花采用正方形的重复排列手法，细致的跌级，韵律感十足的弧形波浪线，创造出丰富的视觉效果。展示区以深色的柜体搭配古铜色不锈钢，衬托出饰品的多姿多彩，呈现华美绚烂的视觉空间。家庭厅以米白色为主色调，配搭浓厚西方挂画，彰显主人高贵格调。主人房背幅摒弃了复杂的肌理和装饰，利用简单的扪皮线条铺贴，细致跌级以对称的手法营造出简约典雅的风情。浴室选用黑檀木饰面配以圣罗兰石，墙身大面积的顶级名贵鱼肚白石，尽显法式半岛风情。

ARMANI

WASHINGTON
LYON
NEW YORK
PARIS
JACK RUSSELL:
Dog Detective
Inspector
Jacques

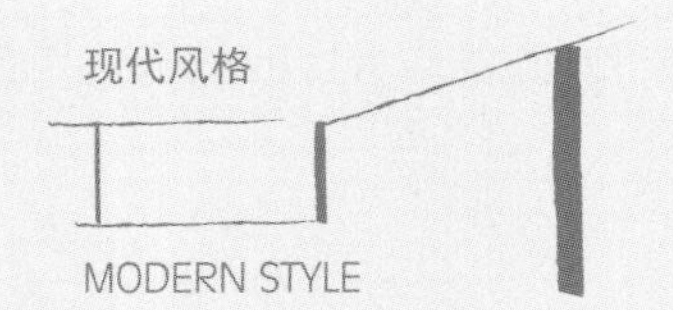

DESIGN CONCEPT 设计理念

This project adheres to the extreme concise spirit of Armani, uses its harmonious and elegant taste on soft decoration, and creates a low-key, luxurious and elegant residence. The living room is injected with design concept of Armani, which emphasizes comfortable, concise and elegant principles. The TV background uses simple and symmetrical hidden bar lights to set off the exquisite white jades, which increases the depth of the space. The designs in the dining room and living room are the same. Dark dining table, elegant light gold upholstery dining chairs and the distinct droplight complement each other. On the other side there is a bar, collocated with high chairs, which enriches the owner's life quality. The family room adopts traditional Chinese sitting pattern. The wood texture paintings collocate with dark color Armani sofa, which creates an Oriental atmosphere. The background of the master bedroom uses regular shutters, which shows stratified changes of lights. The superior Armani silk fabrics and pillows and texture handmade carpets present an elegant and luxurious tone.

AN ELEGANT SPACE

典雅空间

项目名称：新力洲悦示范单位叁	设计公司：奥迅室内设计有限公司	设计师：罗海峰
项目地点：江西南昌	项目面积：600 平方米	摄影师：FOCUS 摄影工作室
主要材料：鳄鱼木饰面、黑色钢琴漆木饰面、清水玉石、黑白根石、雅士白石、墨翠石等		

本方案秉承 Armani 的极端简约精神，利用其和谐蕴藉的高雅品位加以运用在软装装饰上，创造出一所低调奢华的典雅居亭。客厅融入着 Armani 的设计理念，强调舒适、简约典雅的原则。电视机背幅利用精简对称的条形藏光衬托玲珑白玉石，增加空间层次感。餐厅与客厅的设计一脉相承，深色餐台和高雅淡金色的扪皮餐椅，与极具韵味装饰吊灯相互辉映。另一侧设置吧台，配合高脚餐椅，丰富主人生活品质。家庭厅布置遵循传统中式对坐手法，木纹肌理挂画配以浅色系 Armani 沙发，营造出东方情怀的气氛。主人房背幅采用规则的百叶排列方式，展现出灯光的层次变化。Armani 高级的丝质布料与抱枕，肌理纹手工地毯，呈现高雅奢华的格调。

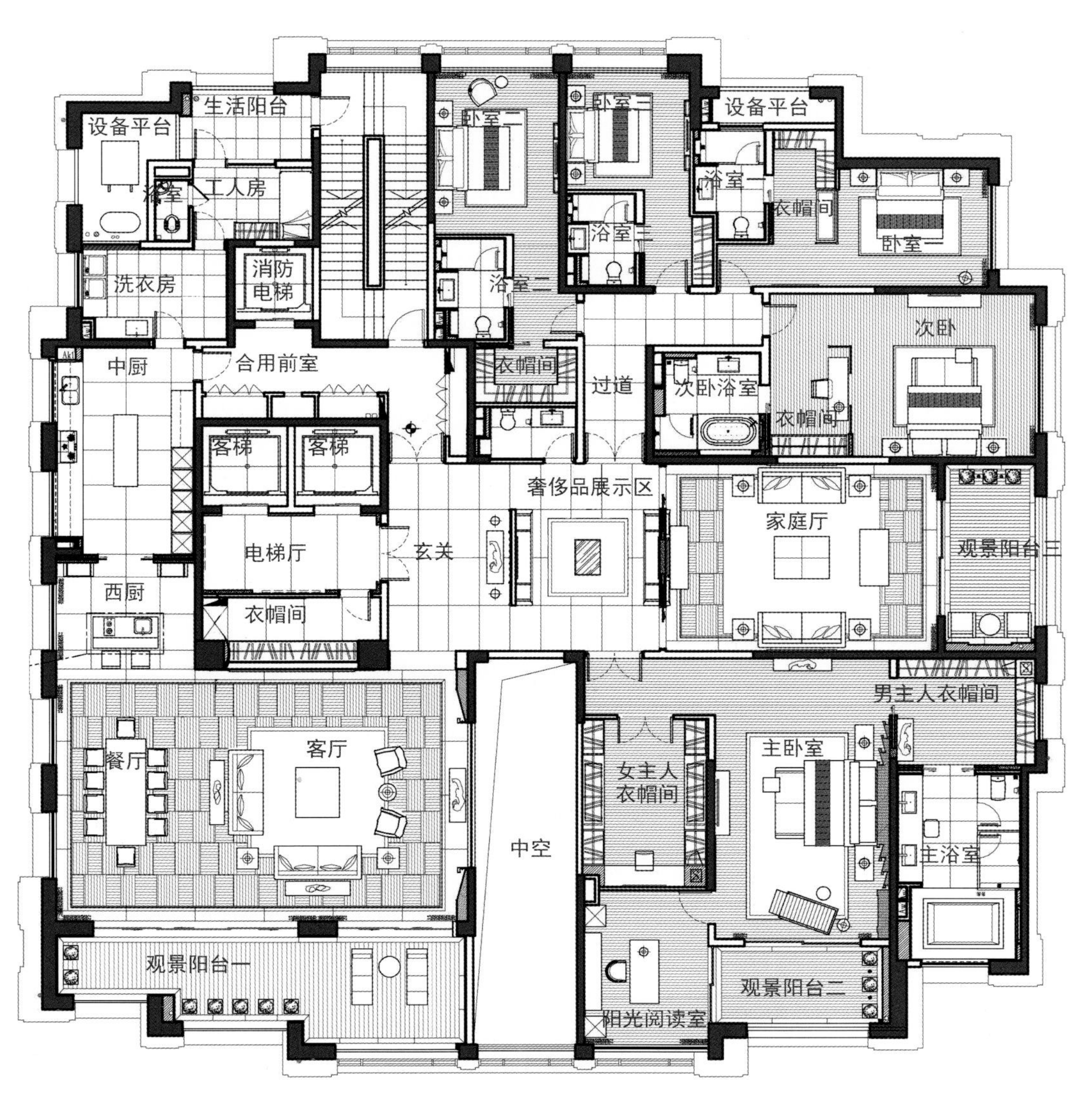

生活阳台
设备平台
浴室
工人房
卧室
卧室
设备平台
浴室
衣帽间
卧室一
浴室三
浴室二
洗衣房
消防
电梯
中厨
合用前室
衣帽间
过道
次卧浴室
衣帽间
次卧
客梯
客梯
奢侈品展示区
家庭厅
观景阳台三
电梯厅
玄关
西厨
衣帽间
男主人衣帽间
餐厅
客厅
中空
女主人
衣帽间
主卧室
主浴室
观景阳台一
观景阳台二
阳光阅读室

DESIGN CONCEPT 设计理念

The house has three layers, whose style is modern simple style which the designer is good at. The double values of space planning and visual communication are presented perfectly in this project.

The seven meters horizontal hall shows decent and direct aesthetics. The size of furniture such as sofa, table and chair is suitable for the living room, which ensures complete functions and makes full use of the space. The background of the sofa is decorated with exquisite concavo-convex lines, which displays the temperament in detail. The furnishings of different sizes on the tea table create visual lightness. The multi mix-match presents a distinct fashion taste. The calm wood background echoes with elaborate furnishings. The style of every space is unified and coherent. Different lines and rhythm spread in the space. The vertical segmentation of the wall presents a tridimensional sense of layering. The position of TV collocates with collecting and displaying cabinets, which is beautiful, neat and practical. The application of decorative materials in the living room echoes with each other. It uses dark background to match with light color furnishings visually, which is harmonious.

SIMPLICITY MANIFESTS DESIGN TEMPERAMENT

简约中方显设计气质

项目名称：万科合肥森林公园馨园叠加样板房　　设计公司：壹舍室内设计（上海）有限公司　　设计师：方磊

参与设计：马永刚、朱庆龙、黄大康、周莹莹　　项目地点：安徽合肥　　项目面积：283 平方米

主要材料：木饰面、石材、壁布、墙纸 、拉丝不锈钢镀黑钛、皮革等

本案上下空间共三层，设计师的风格定位是他最擅长的现代简约风。空间规划和视觉传达，双重的价值在本案中得到了最大呈现。

开间 7 米的横厅展现的是大方、直接的审美。客厅中选择体量与空间适应的沙发、桌椅等家具保证使用功能完善的同时，也让空间显得游刃有余。沙发背景饰以精致的凹凸线条，于细节中见品质。茶几上长短高低错落的饰品造成了视觉的轻盈感，多元化的混搭透出不拘一格的时尚气息。设计采用沉稳的木饰背景，与精心搭配的家居相互呼应，每个空间风格统一，连贯一体。大小分割的线条和韵律在空间中展开，墙体竖线分割，呈现立体层次。电视的摆放结合收纳及展示柜，美观、整齐而又实用。客厅在装饰材质的应用上也是相互呼应的，视觉上用深色背景配浅色家居，整体尽显和谐。

The display technique of furnishings is different from the previous. The height, size and direction are well-arranged, which breaks the inherent mode. Every point of view is a perfect display. The entrance and hallway disperse their kinetonema leading to the living room and dining room separately. Abstract art paintings are the visual focuses of the space. They endow the space with unique art temperament and bring flowing artistic beauty. The design of the dining room continues the style of living room, collocated with vibrant plants, which creates a comfortable dining atmosphere. Because the window of the bathroom in the first floor is beside and above the wash basin, the mirror is designed as mobile hanging type, which can move from right to left and is easy to use without influences on day lighting.

饰品的陈列手法区别以往，高低、大小、横竖不一，错落有致，打破了固有的模式，每个角度都是完美的呈现。入口玄关将动线分散，分别通往客厅与餐厅。抽象的艺术画是整个空间的视觉焦点，画作赋予空间独特的艺术气质，带来充满流动性的艺术美感。餐厅部分设计延续了客厅的风格，搭配生机盎然的绿植，营造出舒适的就餐氛围。一楼卫生间因为窗体位置出于洗手台的上方，为了方便使用，镜子被设计成可移动的吊挂式镜体，可以左右移动，既满足使用，又不影响卫生间的采光。

The master bedroom in the second floor chooses neutral gray and beige and creates elegant atmosphere by furnishings and lamps. The ornaments of dark pillows and carpet and the use of colors and materials deduce a distinct and luxurious temperament. On the layout, the master bathroom is designed with the concept of two separated wash basins. Independent basins with pedestal and full-length mirror make the space transparent and spacious and reflects the living state of the owner. After a busy day, one can relax and enjoy their own time.

二楼主卧依然选择了灰色和米色的中性色调，用装饰品和灯具来营造优雅的气氛。以深色抱枕和搭毯来点缀，用色彩和材质演绎出一种卓有个性的奢华气质。主卫是从布局上看，是两个台盆分开放置的设计理念，独立的柱盆，整面的落地镜，使整个空间通透宽阔。体现了户主的生活状态。在劳累一天后，可以在这里尽情放松，享受完全属于自己的时刻。

In the family room in the basement, the designers partition two spaces by cambered carpet and the layout of furniture, break the stiffness of the space, and create relaxing family atmosphere. The studio in the basement is positioned as photography and travel, and connects with family hall, sunken yard and household dark room. Clean and concise lines are used in every space by the designers. Among the lights, there comes a restrained and capable temperament from the inside out.

地下室的家庭活动室，设计师通过弧形的地毯，从家具的布局和摆放来分割两个空间，打破空间的呆板，营造轻松的家庭式氛围。地下室的工作室空间，以摄影旅游爱好者为定位，连接家庭厅及下沉式庭院和家用暗室。干净简洁的线条被设计师张弛有度的运用到空间中，在灯光线条的交错里，由内而外呈现出一种内敛干练的气质。

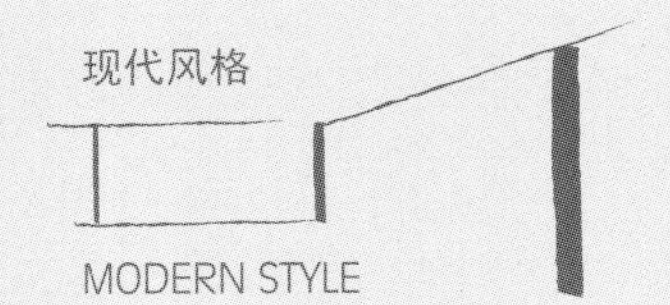

DESIGN CONCEPT 设计理念

The main target people of this project is young people who pursue fashion. Their knowledge and aesthetic perceptions are more open, more diverse and freer. So the young designers use modern blue as the theme, supplemented by gold, to endow the space with colder temperament. The shapes of furniture form decorations and furnishings in different degrees and constitute a powerful "ecological chain". The changes and circulatory complement of colors create a modern space with its own personality. Hard materials, such as stone, metal and wood veneer, combine with soft covers and wallpapers. In addition with fashionable lines from Art Deco, fluent geometrical crafts are presented in more details on the floors and accessories. Get away from the busy work, one can take a thorough rest at this moment.

MODERN METROPOLIS

摩登都市

项目名称：75 m²户型样板房	设计公司：PMG. 伍重室内设计	设计师：梁苏杭、李晴、陈艳萍
项目地点：重庆	项目面积：86 平方米	

本案例的主体面向对象为追求时尚的年轻人，他们知识层面和审美观点更开放、更多元化、更慵懒。所以年轻的设计师在设计时以蓝色摩登为题，金色作为补充，让空间散发出更为冷冽的气质；家具形态则不同程度上形成装饰点缀，构成一个有力的“生态链”。色彩的冷暖交替、循环互补，打造了一个自我风范的摩登空间。材质上运用了石材、金属、木饰面等硬质材质，结合软包、墙纸。Art Deco 的线条时尚也融入其中，流线几何造型的工艺体现在更多的地面和配饰细节上。从繁忙的工作中抽身，这一刻你需要彻底地放松。

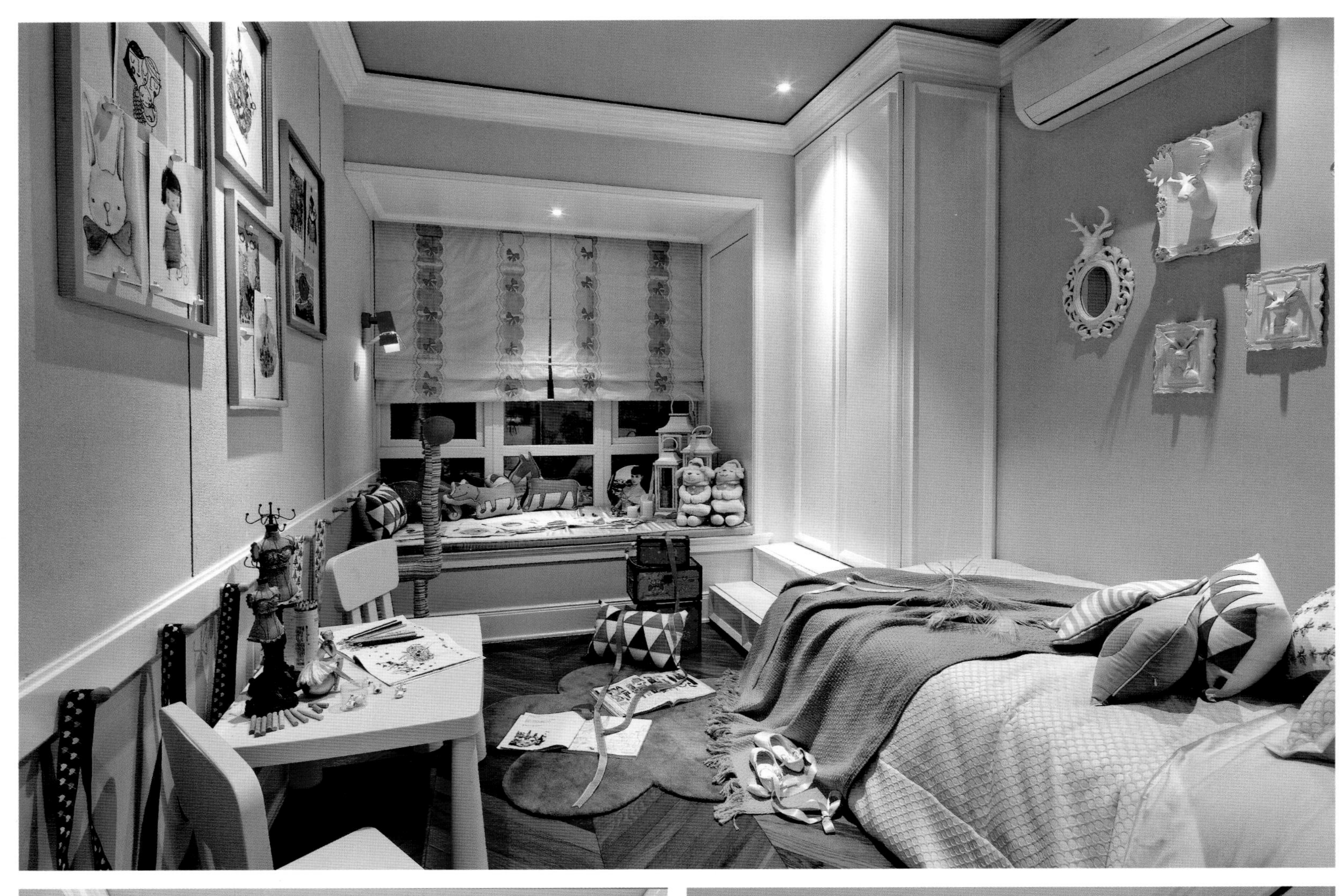

Arletta
SUPERB DESIGNS
ANNIE KELLY

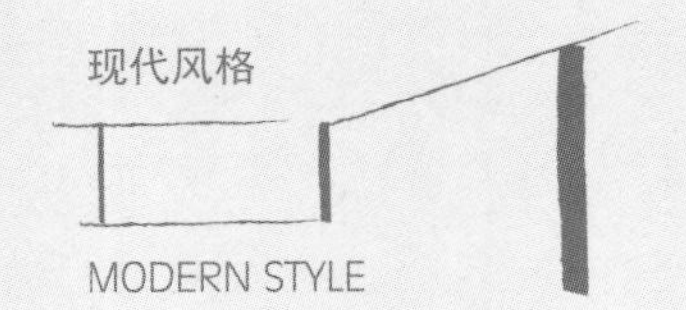

DESIGN CONCEPT 设计理念

When we are talking about design, what are we talking about?

The most amazing thing in the world is like that you cannot find the most accurate definition for it while when you look around, its traces are everywhere. Few people can explain the concept of design. The sunshine pouring through glass windows in the water villa when the belles of the chapelle de ronchamp ring, or the Marseille Apartment with new lifestyle by waking people up from the boring life postwar, are masterpieces by designs. When design pioneer like Mies Van Der Rohe integrates modern aesthetics with our daily lives, people never stop exploring how to combine modern home design with classical aesthetics. The design we talk about is a modern lifestyle, and an attitude of integrating modern aesthetics with classical aesthetics. Luxurious decorations can only bring temporary visual impressions, touching the emotion is truly memorable.

CHARACTERISTIC INTEGRATION OF AESTHETICS

美学的特色融会

项目名称：深圳香山美墅样板间	设计公司：壹舍室内设计（上海）有限公司	设计师：方磊
项目地点：朱庆龙 、刘存森、耿一凡	项目面积：430 平方米	摄影师：罗文
主要材料：不锈钢镀古铜、橡木染色饰面、珍珠贝母、意大利木纹石等		

当我们在谈论设计时，我们谈论的是什么？

世界上最神奇的事情大都如此：你无法为他找到最准确的定义，但当你举目四望，周遭却都是它的印记。很少有人能阐述设计的概念，但当朗香教堂响起的钟声，流水别墅中透过玻璃窗倾泄而来的阳光，或是把人们从战后乏味的生活中唤醒，创造出新的生活方式的马赛公寓，都是设计幻化的杰作。当密斯凡德罗这样的设计先驱，让现代美学融入到我们的日常生活之后，人们就永不停歇的探索如何将现代居室设计与古典美学相结合。我们谈论的设计实则是一种现代的生活方式，一种将现代美学与古典美学融会贯通的态度。华丽布置只能带来视觉上一时的印象，触动情感才真正令人难忘。

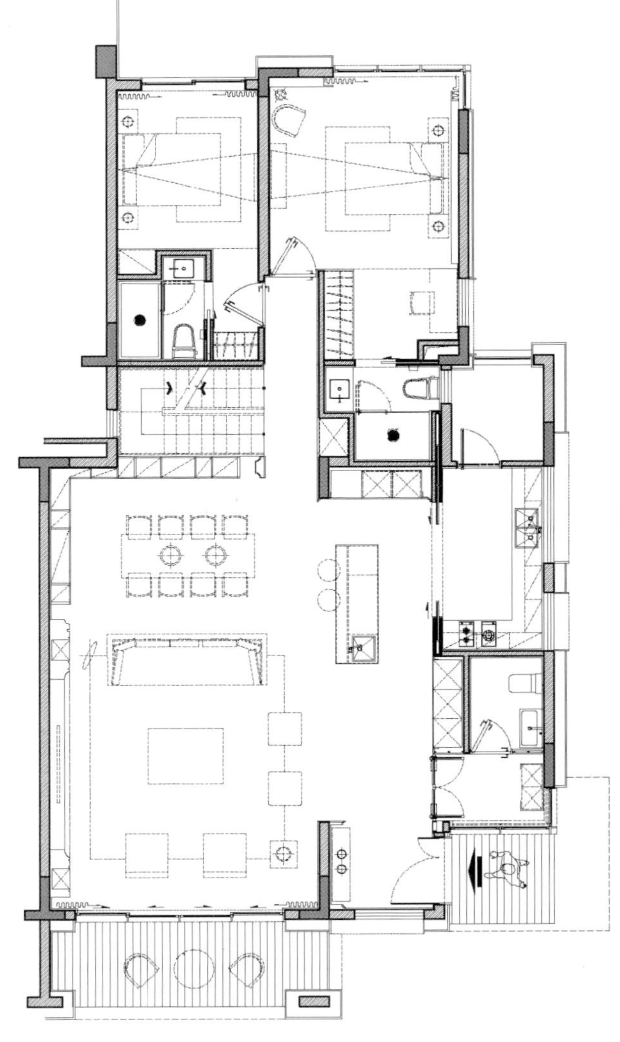

The project is located in Shenzhen, "City of Design", which connects with Hong Kong by mountains and rivers. It uses new methods to deconstruct classical style, continues concise spirits, and promotes and discovers lifestyle and modern aesthetics starting from contemporary life. The theme is to salute to design pioneers, aiming to create contemporary residence in people's mind.

The functional layout tries to create more possibilities in the space. The platform of the open kitchen in the first floor becomes a joint connecting dining room with kitchen, which continues the dining function, strengthens the exchanges between kitchen and dining room and effectively connects kitchen, dining room and living room. The kinetonema in the second floor is set clearly with the family hall as the center, full bedroom on suite at two sides. The viewing balcony in the master bedroom makes the view wider. The reasonable functional areas and accurate partitions in the underground meet various needs daily life, and provide every family member with more independent space.

本案位于与香港山水相连的 "设计之都"深圳，用新的手法来解构古典，设计上延续了简约的精神，从当代生活出发，对生活方式及现代美学提升和发现，向设计先驱们致敬是本案设计的主题，旨在打造人们心目中的当代居所。

功能布局试图营造出更多空间的可能性，一层开放式厨房的中岛台，成为餐厅与厨房的连接点，既延续了用餐功能，又加强了厨房和餐厅的交流，有效将厨房与餐厅、客厅相连。二层的空间动线划分明确，以家庭室为中心，两侧分别设计全套房卧室，主卧置入的观景阳台，视野范围更加开敞。地下一层合理的功能区间以及精准的分区让家庭生活的各种需要得到满足，更让家庭成员拥有各自独立的活动空间。

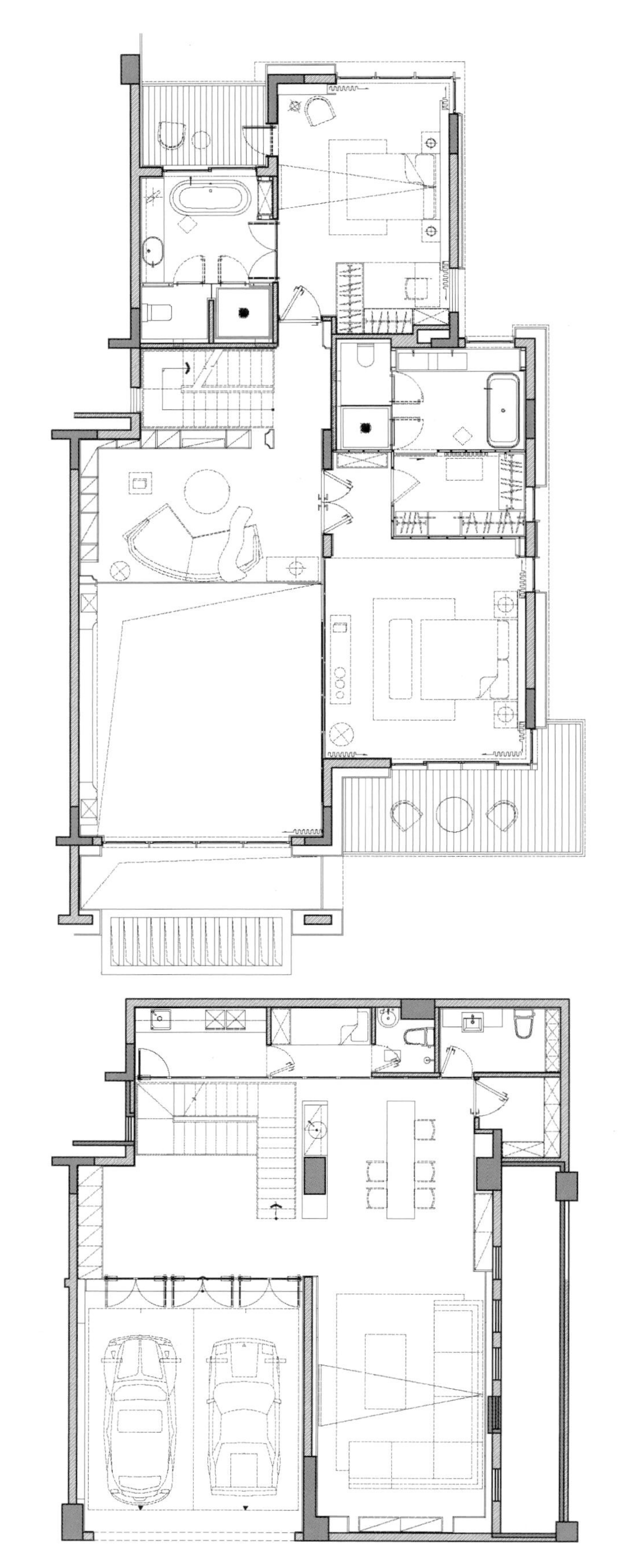

在一层近6米的挑高客厅里，我们试着寻找大师的足迹，精致黑色皮革电视背景墙与粗犷大理石天然纹理对比，明快的线条感以及在材质搭配中注入有变化的细节，简练而制作精致的细部，灵活的空间动线，无一不在向经典致敬。

梭罗（Henry David Thoreau）曾说过，我的屋子里有三张椅子，独坐时用一张，交友时用两张，社交时用三张。每个人的生命中都应该拥有一把扶手椅，它或是古典优雅，或是现代摩登，如同遁入温柔乡，在这里，你可以做最奢侈的事情——享受时光的美好。

现代家居室内设计的趋势，正在寻觅恢复和重塑优雅，客厅整体沉稳大气的皮质沙发搭配颜色轻快的躺椅，宁静且多功能的家居配饰，使空间洋溢轻奢的态度。

餐厅白色烤漆的柜体为空间增添了时尚现代感，石材边框的细部线条处理是在古典造型下的简化，在细节处赋予怀旧的内涵。线条干练的餐椅造型极具现代感，大气水晶吊灯，祛除了传统复杂的造型，为利落的餐厅增添了几许复古气息，视觉上不显冗杂，又别具匠心，在整体空间里，将古典元素用现代方式演绎，在环境中酝酿出当代的人文精神。

二层首先是家庭室，相比客厅的稳重大气，家庭室相对休闲轻松。米白色的布艺沙发搭配浅色大理石茶几，使空间更加温馨雅致。主卧空间沿袭了客厅的风格，在材质运用上做出不同层次，以深棕色的木饰面为基调，中色调的皮革搭配浅色的贝壳马赛克，白色烤漆线条，不同层次和肌理的材质，都有着精致的做工。床上的紫色丝绒抱枕和床尾凳，则为空间带来高贵复古的气息和触感。开敞式的衣帽间，不仅是主卧到主卫的空间过渡，在这里也能同样反映出主人追求细节，洗练雅致的品位。

地下一层的设计则更加注重功能性，不同功能区域的划分使地下空间更具娱乐性。红酒雪茄区、影视区、吧台区一应俱全，硬装没有多余的装饰，整体统一的色调一气呵成。利用家具不同陈列方式来划分功能，低调中透显静享生活的感悟。

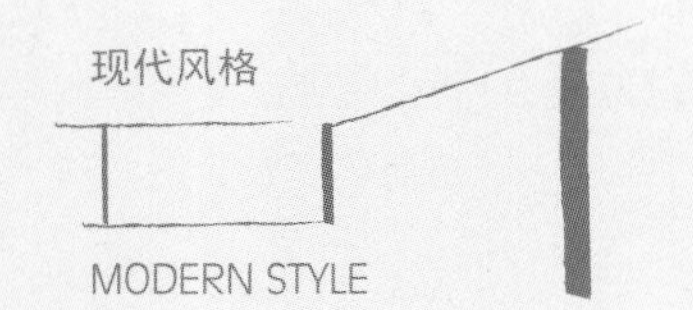

DESIGN CONCEPT 设计理念

Ensenberg, a famous Germany reporter once said, "In the rapidly developed future, the definition of luxury is no longer inessentials of life, such as famous cars, gold watches, limited perfume, it is basic conditions of peaceful time, enough water and air." Luxurious life is not material satisfaction but the possession of scarce resources which is really close to authentic luxurious life. So the top real estate which is closest to scarce resources has already become the new favorite of the rich.

德国著名记者恩森贝格曾说，"在高速的未来发展中，奢侈品的界定将不再是名车、金表或限量香水等生活的非必需品，而是拥有宁静的时光、足够的水和空气等基本条件。"奢华生活如今已经不仅仅是一种物质上的满足，对稀缺资源的占有才是与真正的奢华生活靠拢，于是与稀缺资源最紧密的顶级物业俨然已经成为富豪新宠。

GIVING THE SPACE THE LARGEST LUXURY

给予空间最大的奢华

项目名称：苏河湾滨水别墅	设计公司：壹舍室内设计（上海）有限公司	设计师：方磊	参与设计：马永刚、赵晔琪、李丽娜
项目地点：上海	项目面积：1500 平方米	摄影师：Peter	
主要材料：木皮染色、拉丝玫瑰金纳米、夹丝玻璃、茶镜、淼漫灰等			

This project is located in Shanghai, faces Garden Bridge of Shanghai on the east, connects in one vein with The Bund, faces with Lujiazui across the river, and is near the most international business circle in Shanghai, such as People's Square, Nanjing Road and Huaihai Road. The architecture of this villa integrates characteristic elements in old Shanghai, and combines Oriental and Western culture with distinct tastes, which makes its delicacy shuttle back and forth in times and becomes charming with time going by. The project of 1500 square meters has seven floors, four floors underground and three floors on the ground, positioning as simple unban style. The complicated and diverse space form endows the space with advantages to design various functional areas to satisfy the owner's diverse life needs and interests. The designers are honored to take part in the interior design of this project. The designs fully concern the combination of indoor and outdoor environment from the planar function layout to reach spiritual pursuit of contemporary life and pursue harmonious resonance of culture, art and life.

本案位于上海，项目东临外白渡桥，与外滩一脉相连，与陆家嘴隔江相望，紧邻人民广场、南京路、淮海路等上海最具国际化氛围的商圈。苏河湾滨水别墅其建筑融合了老上海特有的元素，结合东西，独具品位，让它的精致在时光中穿梭，任时光变迁而愈发迷人。该项目共 1500 平方米，地上四层地下三层，定位为简约都市风。相对复杂多样的空间形式赋予了它有利于设计各种功能区域，以满足主人多方面的生活需求和兴趣爱好。壹舍设计顾问有幸参与了滨水别墅的室内设计，设计从平面功能布局就充分考虑到与室外的环境相结合，以达到当代生活的精神诉求，追求文化，艺术与生活的和谐共鸣。

Remington

一楼进入室内，是个全开放式的活动休闲空间，通过天花造型与地面材质变化来限定各区域。空间以白色为主调，让整个空间氛围轻松融洽，墙面古铜色金属的点缀也为室内典雅的气质起到衬托的作用。在采光庭院核心筒的墙面上加了一片绿色植栽，将绿意延伸进室内空间。

上到二楼钢琴区域，是个双层挑空结构，彰显大户型的不凡气势，设计风格于现代都市中不乏艺术品位。客厅与餐厅相贯通，分区明确，自由通透，强化了客厅与餐厅宽敞的视觉感。厨房的空间布置较为丰富多变，设计有强调休闲趣味的吧台。通过宽大的落地窗，室外的景观尽收眼底，不论是白天的绿草茵茵，还是晚上的灯光闪烁，都能让人产生室外互为一体的惬意感受。家具中的古铜金属与墙面的金属交相呼应，让整体空间尽享精致的品位。

三楼起居室的暖色沙发和带纹理的地毯彰显空间的华贵气质，背景波浪形的石材与凹凸有序的装饰线，给人以视觉冲击力与设计的美感。

主卧位于顶层四楼，空间以白色为主，以简约的线条和立面呈现，不失饱满，尽力凸显空间的通透性与连贯性。家纺布艺也是运用相近色系来迎合整体卧室的风格，给人和谐统一的整体感，而少量的暖色灯光的加入，让冷色调的空间更显温馨浪漫。地铺深浅相间的马毛地毯，既加深了空间的层次感，又让整体空间的气质得到升华。整个空间是通过床背景的小矮墙来划分区域，可以在床尾的沙发上沐浴着阳光，惬意地坐着聊聊天，矮墙后方可作为读书思考的区域，一侧的水吧区可以满足主人的闲情雅致。主卧、梳妆间与更衣间在空间上实现了相互连通，方便生活起居。透过夹丝玻璃移门是主卧卫生间，通透宽敞，采光自然，给人身心放松的生活享受。

负一楼为收藏展示区兼具书房的功能，空间以木色为主，色彩相对较深，适于营造安静的阅读思考气氛。书房通过旋转门与水吧区隔开，让这部分空间有多种空间形式。雪茄品酒区以拼花石材为背景，弱化其余墙面的装饰，主次分明。搭配布艺沙发和皮质单椅提升会客空间的档次，一角古朴高雅的留声机，点缀出富丽堂皇的气质，独特而又高贵的设计，是一种彰显个性、品味成功与生活的释怀，是一种只可意会不可言传的独特感觉。影视厅墙面是清新的绿色软包，简约的造型，可以在此放松心情享受视觉盛宴。

SPA 区位于负二楼，墙面石材通过光面与毛面及分割大小，结合灯光效果，让墙面富有变化。简练的设计细节突显出空间的精致，主人可以在这里让疲惫的身心得到释放。

整个设计摒弃了过分浮华的堆砌，通过清晰简明的线条与家具组合，带来的不仅是华贵的质感，而且传递出了浪漫的生活格调，一种对品质生活的享受姿态。

美式风格

AMERICAN STYLE

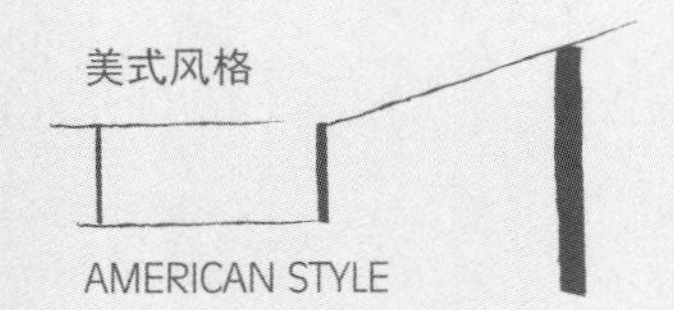

DESIGN CONCEPT 设计理念

"Inheritance of residence" is one of the words which the designer uses to position this single-family villa. It is located in the villa residential area near the Niushou Mountain in Nanjing. The area is built around the mountains and keeps the sloping fields with beautiful environment and quality air, which is suitable to live for a long time. So the designer makes it a residence inheriting American style. After understanding the family formation of the owner, the designer makes a series of transformations, which meets the owner's functional needs and endows the space with American leisure and retro feelings as possible. The American red cherry wood throughout the villa set a sedate tone for the space as if in the late autumn, bringing people comfortable and cozy enjoyment.

"传承式居所"这是设计师喜欢拿来定位独栋别墅的词汇之一。本案位于南京牛首山附近的别墅住宅小区，小区依山而建，基本保留了坡地特色，周围环境优美，空气质量优良，适宜长期居住，所以设计师将它设计为传承美式风格的居所。在了解业主家庭结构后，设计师做了一系列的结构改造，既满足业主使用需求，也尽可能让整个空间更具有美式休闲、复古的情调。贯穿整个别墅的美国红樱桃原木奠定了空间沉稳的色调，如深秋的季节，带给人舒适、惬意的喜悦。

SENSE OF AUTUMN
秋意浓

项目名称：复地朗香 267 栋	设计公司：南京冯振勇室内设计事务所	设计师：冯振勇
项目地点：江苏南京	项目面积：530 平方米	
主要材料：红樱桃原木、精工玉石、天然大理石等		

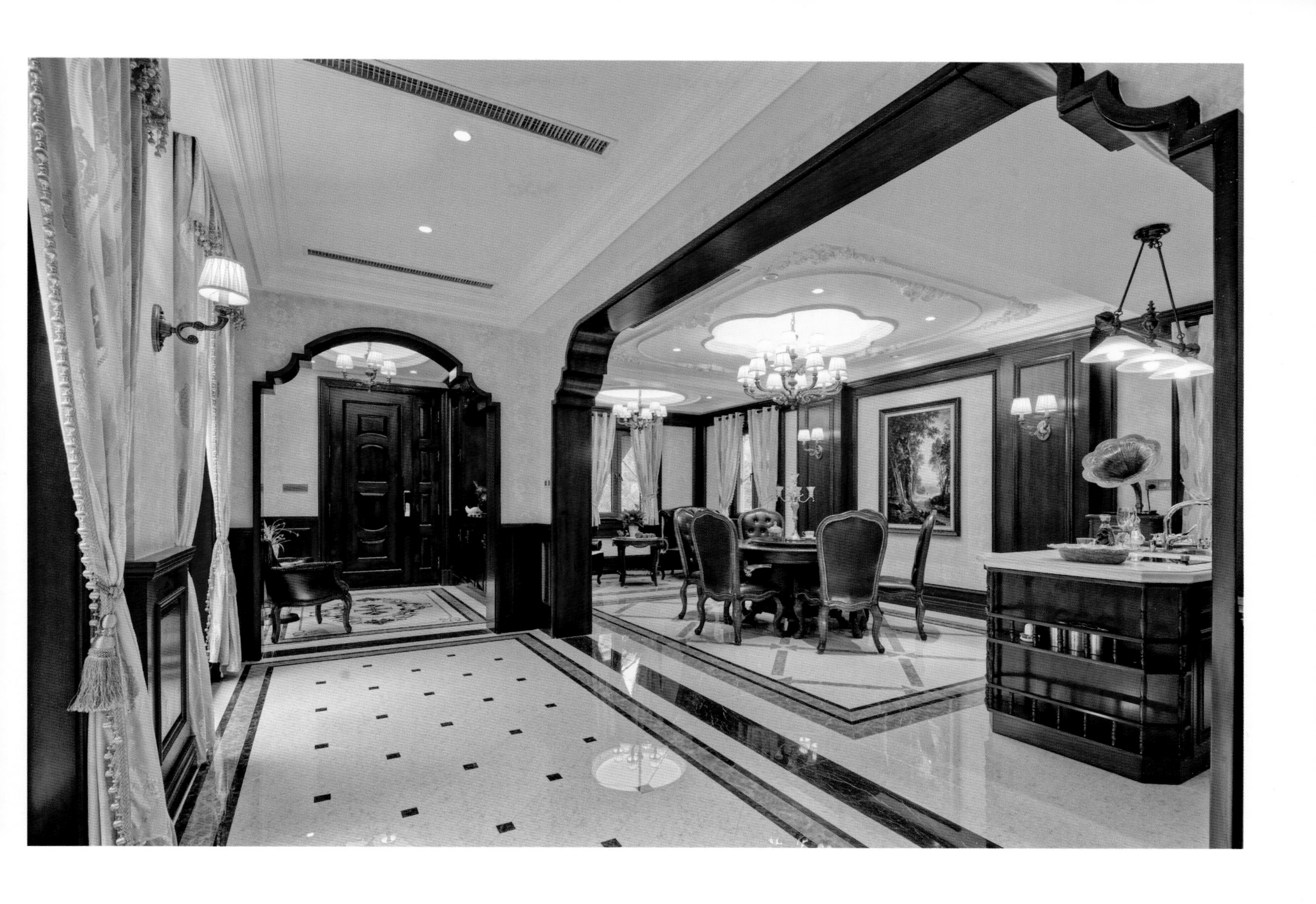

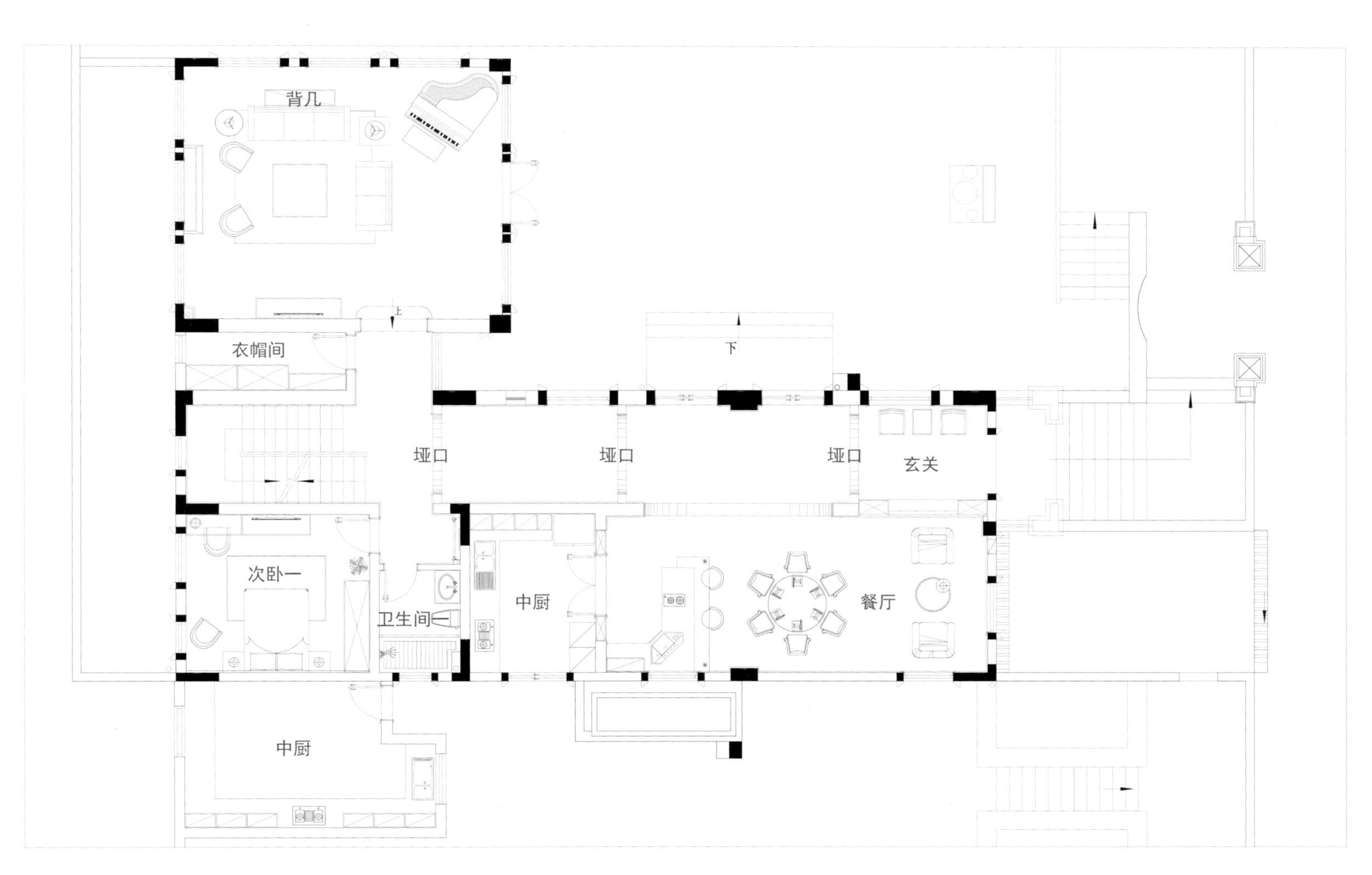
背几
衣帽间
垭口
垭口
垭口
玄关
下
上
次卧一
卫生间
中厨
餐厅
中厨

MANIXUAN
12
WHISKY

"IT'S ALL ABOUT FINDING THE RIGHT CINEMATIC LOOK"
PLAMEN CVETANOV
STONE
WWW.THECINEMATIC.NET
videohive
AVATAR
VIRTUAL RACER

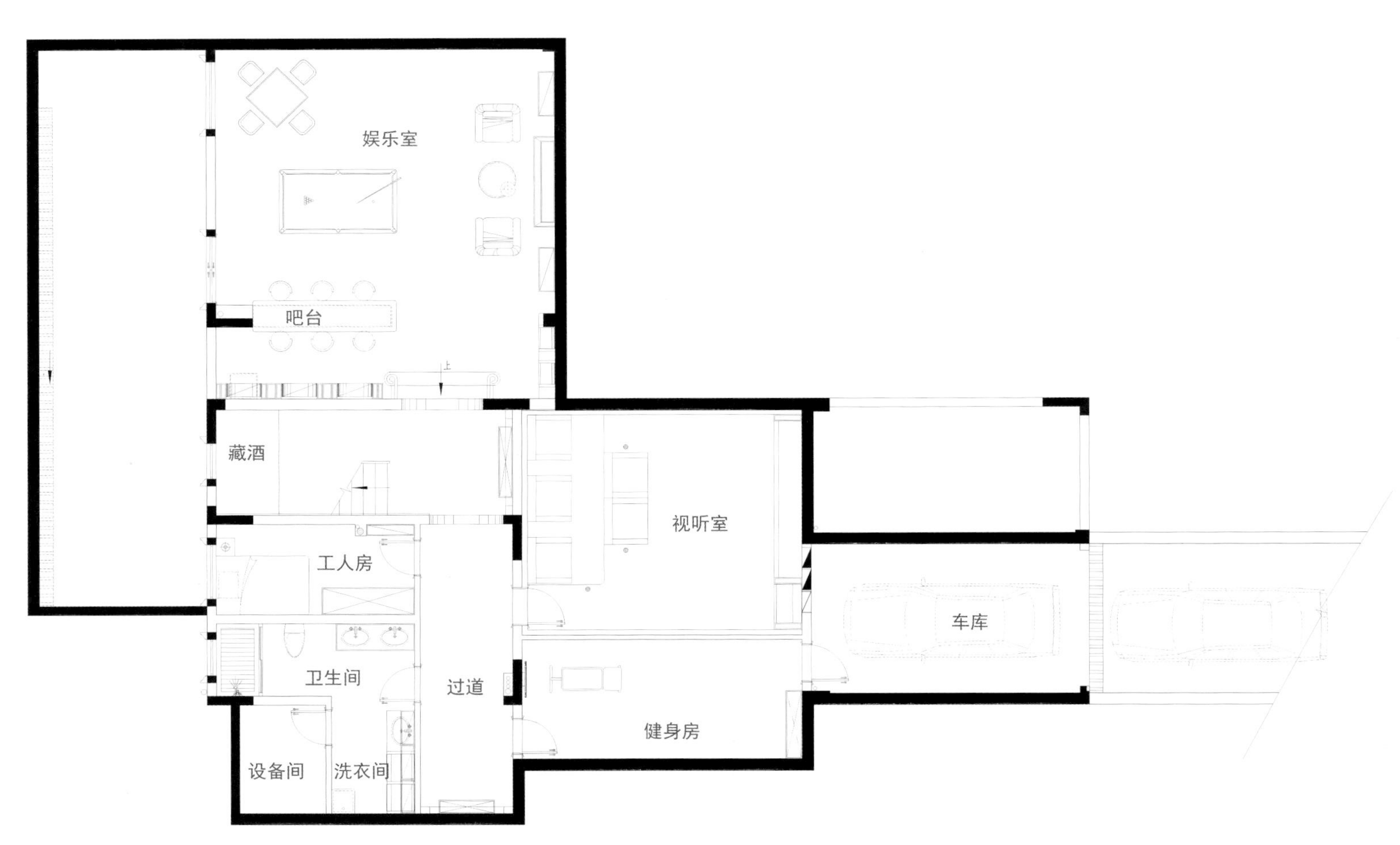
娱乐室
吧台
上
藏酒
视听室
工人房
车库
卫生间
过道
健身房
设备间
洗衣间

ROCK
ROCK
BAR
VINTAGE
POSTER
paris
happines

YACHT
CLUB

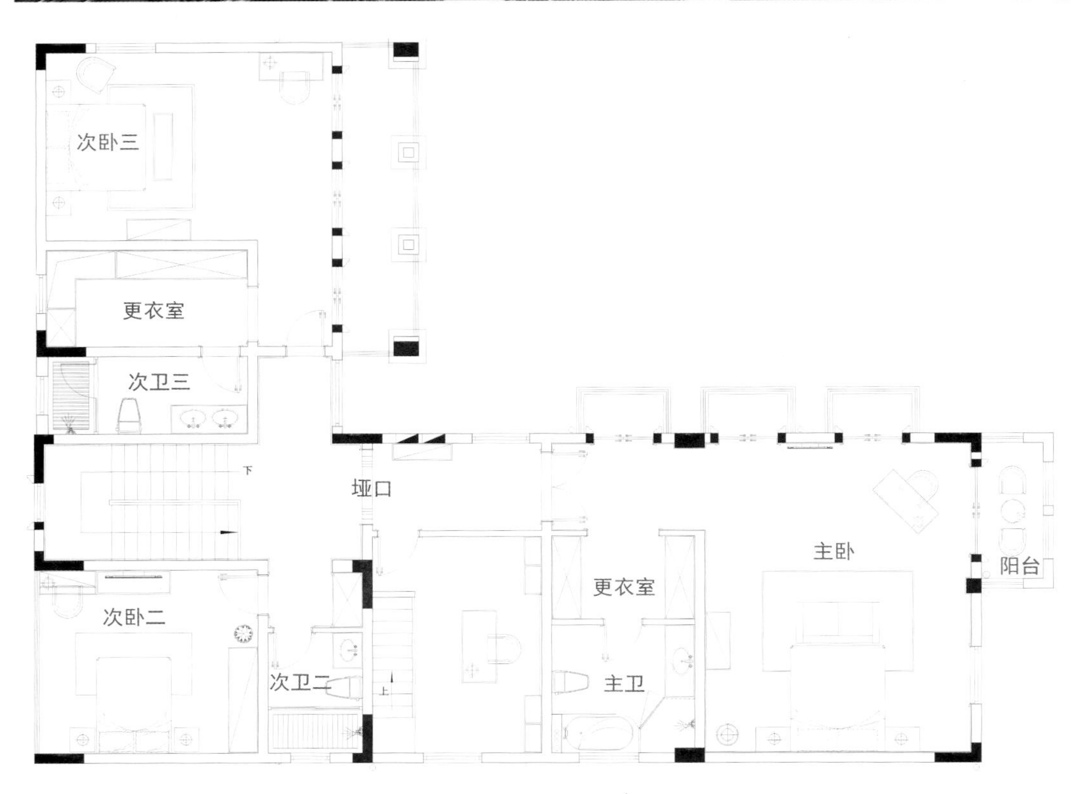
次卧三
更衣室
次卫三
垭口
主卧
阳台
更衣室
次卧二
次卫二
主卫

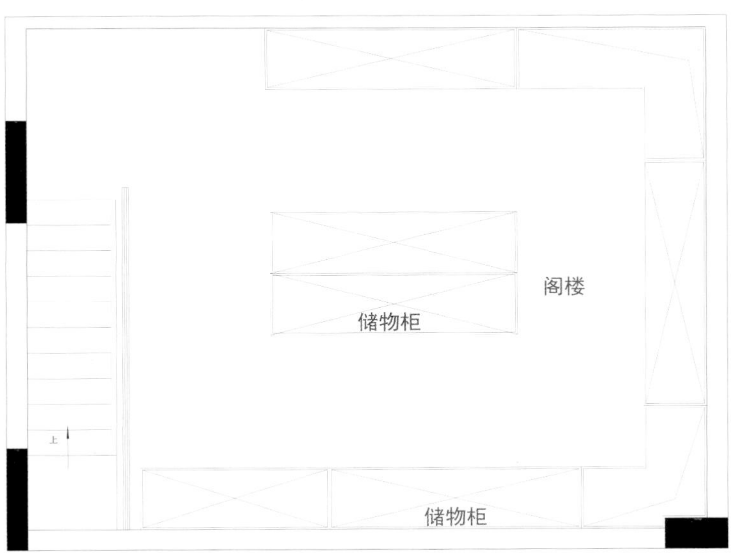
阁楼
储物柜
储物柜
上

DESIGN CONCEPT 设计理念

A cup of Starbuck coffee, a jazz song and an iPad, one can feel the romantic aesthetics, delicacy and elegance. We try hard to find beautiful things and notice the dribs and drabs of life. In fact, the most authentic and enjoyable thing is not self flaunt or decoration in the crowd but individual experience at home alone.

This is an American neo-classical style villa, whose architectural exterior is typical American style while the interior pays more attention to cultural and historical inclusiveness and the depth of the space design. Its essence is from classicism, which is neither archaistic nor retro. American style pursues a romantic charm.

一杯星巴克，一首爵士乐，一台 IPad，罗曼蒂克式的唯美精致而优雅，我们努力发现美好的东西，留意生活上的点点滴滴。其实，最真实写意的，不是在人群中的自我标榜和装饰，而是能独自在家的个人体会。

这是一个美式新古典主义的别墅设计，其建筑外观以典型的美式风格为特色，而室内设计更关注其文化和历史的包容性，以及空间设计上的深度享受。它的精华来自于古典主义，但不是仿古，也不是复古，美式风格追求的是一种神韵。

AMERICAN LUXURIOUS MANSION
美式奢华大宅

项目名称：公园 1903	设计公司：昆明中策装饰集团有限公司	设计师：陈相和
项目地点：云南昆明	项目面积：500 平方米	
主要材料：木饰面、大理石、水晶等		

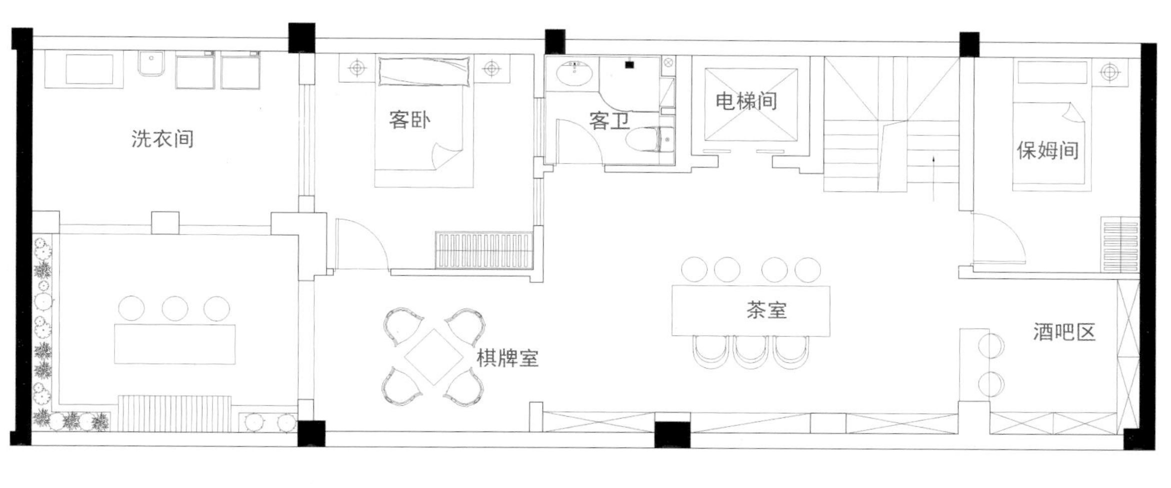
洗衣间
客卧
客卫
电梯间
保姆间
茶室
棋牌室
酒吧区

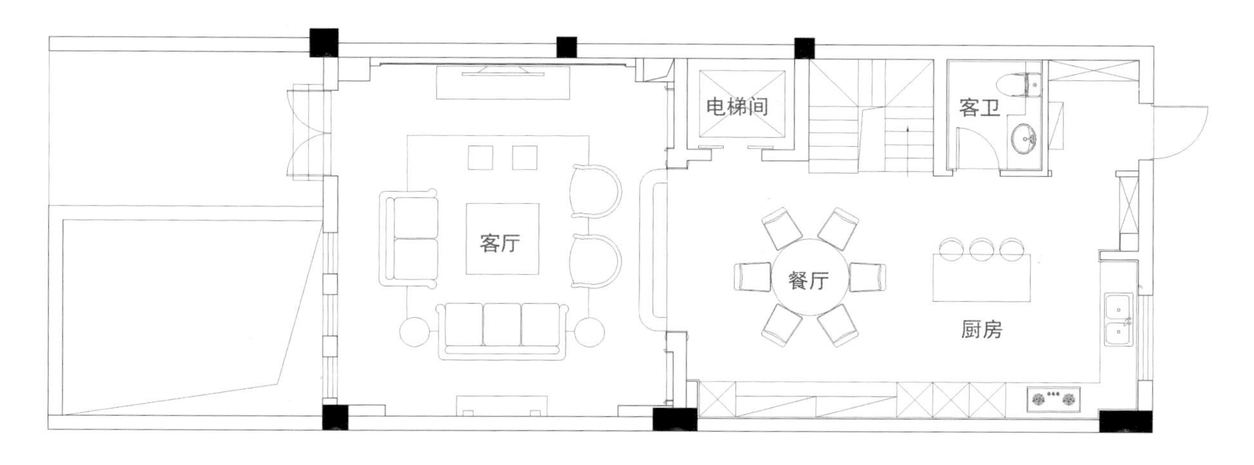
电梯间
客卫
客厅
餐厅
厨房

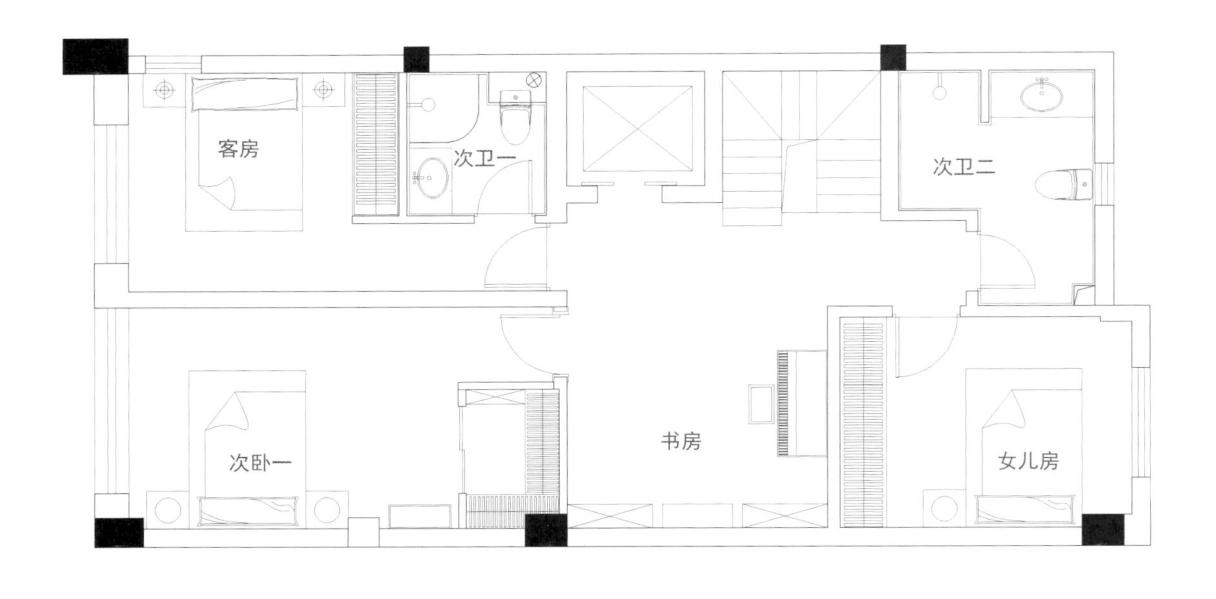
客房
次卫一
次卫二
书房
次卧一
女儿房

The whole space is transparent and bright. As the reception area, the living room is concise and lively, which is more glamorous compared with other spaces. Large areas of carpet collocate with elegant and comfortable American sofa, in addition with the beautiful French windows, which is magnificent and luxurious. The designer skillfully uses technique of expression in American style, stresses living quality, and creates a low-key and luxurious space. The color collocation is also suitable with designer's ingenuity. The tone in the children's room is fresh, clean, elegant and distinct. The color in the elder's room matches dark coffee with comfortable light warm color, which is sedate and indifferent. The decorative paintings make the finishing points, which makes the space immerse in the strong artistic atmosphere.

This is life without luxury everywhere and ornaments of diamond or gold. Let life return to nature, let the soul return to nobility, and let art pervade all around.

整个空间通透且明亮，客厅作为待客区域，简洁、明快，装修较其他空间显得更加光鲜。大面积的地毯搭配造型优雅、舒适的美式沙发，再加上一侧华美的落地窗帘，尽显大气、华贵。设计师巧妙运用美式风格的表现手法，既注重起居品质，又不过分张扬，成就了一室的低调与奢华。在色调的选择搭配上也颇为用心，儿童房的色调清新纯净、淡雅别致；老人房则以深咖色搭配令人倍感舒适的浅暖色系，尽显稳重与淡然。同时装饰画很好地起了画龙点睛的作用，让整个空间沉浸在浓郁的艺术氛围中。

这就是生活，不需要处处摆出奢侈品，不需要钻石、黄金的点缀。让生活回归自然，让心灵回归高尚，让艺术弥漫四周。

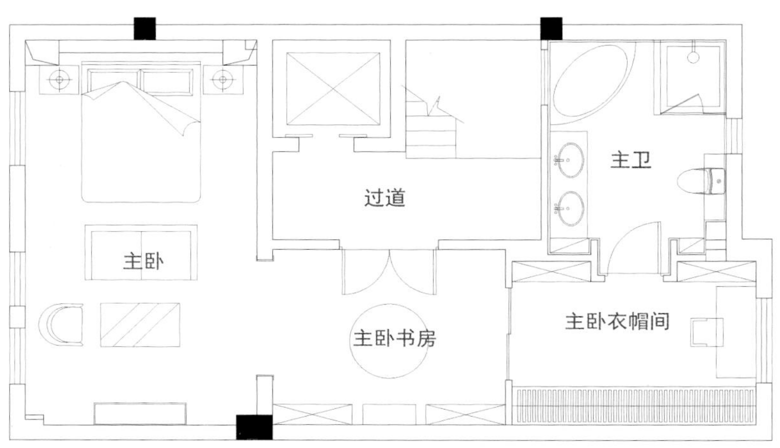
主卧
过道
主卫
主卧书房
主卧衣帽间

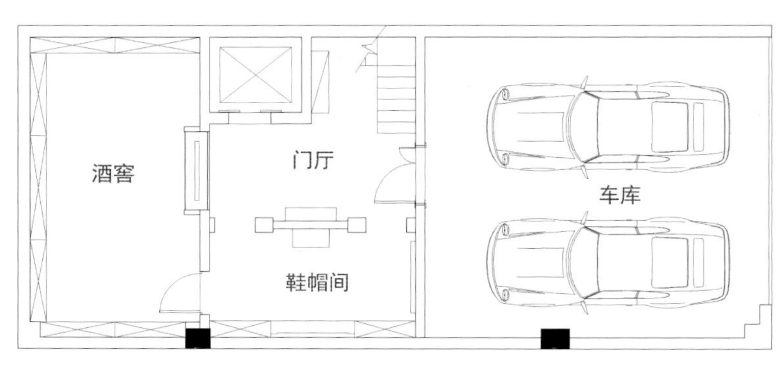
酒窖
门厅
鞋帽间
车库

DESIGN CONCEPT 设计理念

The whole project outlines cultural and texture design taste of American style. As a space for modern people, its emphasis lies on using complete space to create superior and high taste life. The designers make use of textures of materials to create flexible and excellent domain and endow the lines with contemporary taste.

The living room uses embedded plaster ceiling, collocated with exquisite crystal droplight, creating a nostalgic and romantic feeling. The floor is outlined by marbles, which is bright and makes the space partition clearer. The paintings in the dining room and living room choose architectural paintings with better depth of filed, which stretches the depth of the space and is the convergence of art. The designers expect collisions between foreign culture and modern culture with historic taste, which makes the space magnificent and luxurious.

FREE AMERICAN STYLE RESIDENCE

自在美式居

项目名称：南昌绿地悦公馆样板房	设计公司：上海飞视装饰设计工程有限公司	设计师：张力、李姝
项目地点：江西南昌	项目面积：270 平方米	摄影师：金选民
主要材料：大理石、木饰面、壁纸、木地板等		

本案从整体上勾勒美式人文质感的设计品味，作为现代人的使用空间，其设计重点是利用完整的空间区域来缔造优越及高品味的生活。设计师利用材质的肌理与质感，构造出灵动优秀之场域，让线条展现也融入当代品味。

客厅天花板采用嵌入式石膏吊顶，搭配精致的水晶吊灯，营造出一种怀旧、浪漫的感觉。以大理石作为地面装饰的勾勒，光泽感极佳，使空间区域划分也更加清晰。餐客厅的挂画选取的建筑挂画景深极佳，在拉伸空间的纵深感的同时也是艺术的交汇。设计师希望将海外文化与现代文化的精神相互碰撞，富有历史的气息，让整个空间变得大气奢华。

The master bedroom is a better place for gentleman without flatulent decorations or colors. All attitudes towards taste are outlined by quiet and soft black, gray and gold. The soft curtains, corners of hard decoration elements and furniture and texture details can trace back to the glories of American elegance.

texture through every aspect. These elements cater to the needs of people's lifestyle nowadays, namely cultural sense, noble sense, free and emotional senses and lost in the time tunnel.

主卧是鲜明的绅士之居，没有浮夸的装饰或色彩，一切关于品味的态度均在平静、温和的黑、灰、金色调中被勾勒。在柔和的窗帘、硬装元素和家具的一些边角、纹理细节，能追溯到美式优雅的点滴光辉。

设计师从各个方面给予这套居所新鲜蓬勃的生命活力和本质化美式生活的质感，通过这些元素也正好迎合了时下人们对生活方式的需求，即：有文化感、有贵气感，还不能缺乏自在感与情调感，让人迷失在时光隧道中。

FOUR SEASONS

DESIGN CONCEPT 设计理念

It is said that years of precipitations are essences, which perfectly describes the owner of this project. The love to red wine and family of the couple who works on Italian red wine business and helps each other when in humble circumstances provide best inspirations to the designs. With a cup of tasty wine, one can enjoy the peaceful time. The 350 square meters space not only embodies the designs, but also carries the emotions of the family of three and the belonging of years. Fabric sofas and green plants collocate with light color soft decorations, making the space gentle and tender. Exquisite furniture is carefully carved. The texture of parquet and carved gold makes the residence leave a meticulous impression on people from the simplicity to complicity and from the whole to partial. The designers keep the traditional decorative traces and cultural connotations, and abandon unduly complicated texture and decoration, which reflects an elegant, tranquil, warm and gentle atmosphere.

BELONGING AFTER YEARS OF PRECIPITATIONS

岁月沉淀下的归处

项目名称：南京玛斯兰德　设计公司：南京陈熠室内定制设计事务所　主案设计师：陈熠　参与设计师：余杨

项目地点：江苏南京　项目面积：350 平方米　摄影师：Ingallery

主要材料：乳胶漆、大理石、木地板等

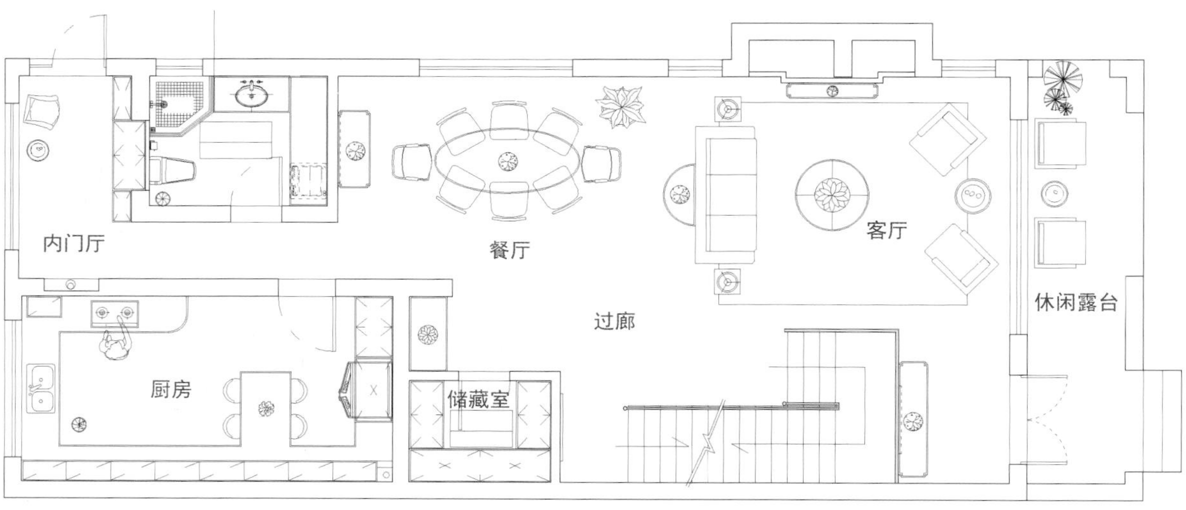
内门厅
餐厅
客厅
过廊
休闲露台
厨房
储藏室

有人说岁月沉淀下来的都是精华，这句话用来形容本案的业主最好不过了。这对从事意大利红酒生意、相濡以沫的夫妻对红酒的热爱以及对家的情感给我们的设计提供了最好的灵感。葡萄美酒，时光静好，这个 350 平方米的空间里体现的不仅是我们的设计，它更是承载着三口之家的情感和岁月的归处。布艺沙发、绿色植物配以淡色柔美的软装，让整个空间柔和淡然。精致的家具，精雕细琢，镶花刻金的质感使得整个居住空间从简单到繁杂、从整体到局部都给人一丝不苟的印象。我们保留了传统装饰的痕迹与文化底蕴，同时又摒弃了过于复杂的肌理和装饰，体现一种高雅、宁静、温馨、柔和的氛围。

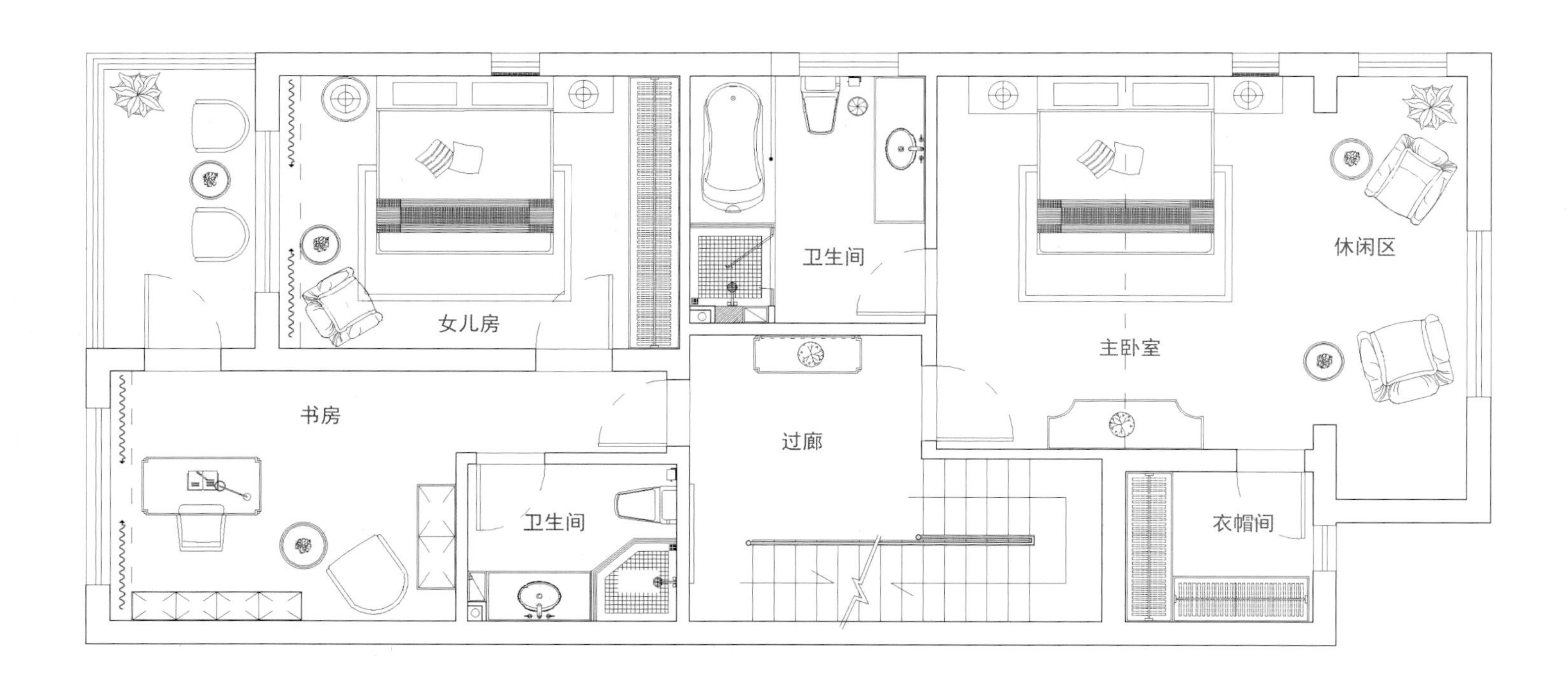

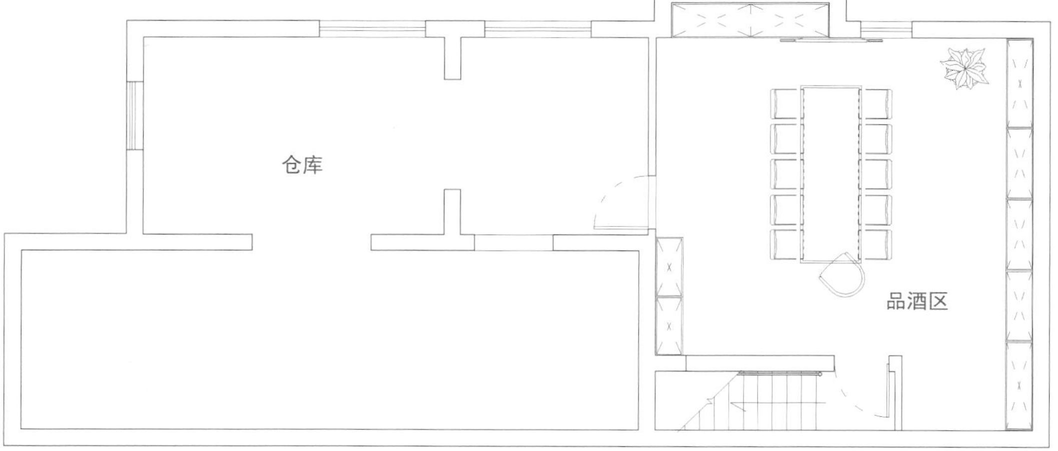
仓库
品酒区

DESIGN CONCEPT 设计理念

This project is located in the intersection of Pushang avenue and Wulongjiang avenue, having rare river scenery and resource from Wulongjiang. The high geographic coordinate redefines the new height of life quality in Minhou County. The plane layout uses counterpoint relationship of space axes to better display main spaces in the house. The elevation treatment breaks the counterpoint relationship by space axes. The collocation of soft decoration makes the space more relaxing. The natural tones, such as creamy white, gray and coffee, largely decrease the color purity. The partially used high saturation colors brighten the space. The displaying tones use popular colors in the spring and summer of the year 2016, such as grayish blue, amber green and Hermes orange. The retro American style collides with concise modern style in the space, which is fun.

RIVERS AND MOUNTAINS
群升江山

项目名称：群升江山城 804 户型复式样板房设计　　设计公司：深圳市万有引力室内设计有限公司　　设计师：白丽雪

项目地点：福建福州　　项目面积：200 平方米　　摄影师：大斌

主要材料：欧亚木纹、土耳其灰洞石、巴萨灰石材、雅柏灰石材、树瘤木饰面、橡木饰面、浮雕金刚柚木地板、墙纸、皮革、不锈钢、瓷砖、马塞克等

What's
cooking?

What's
cooking?

What's cooking?

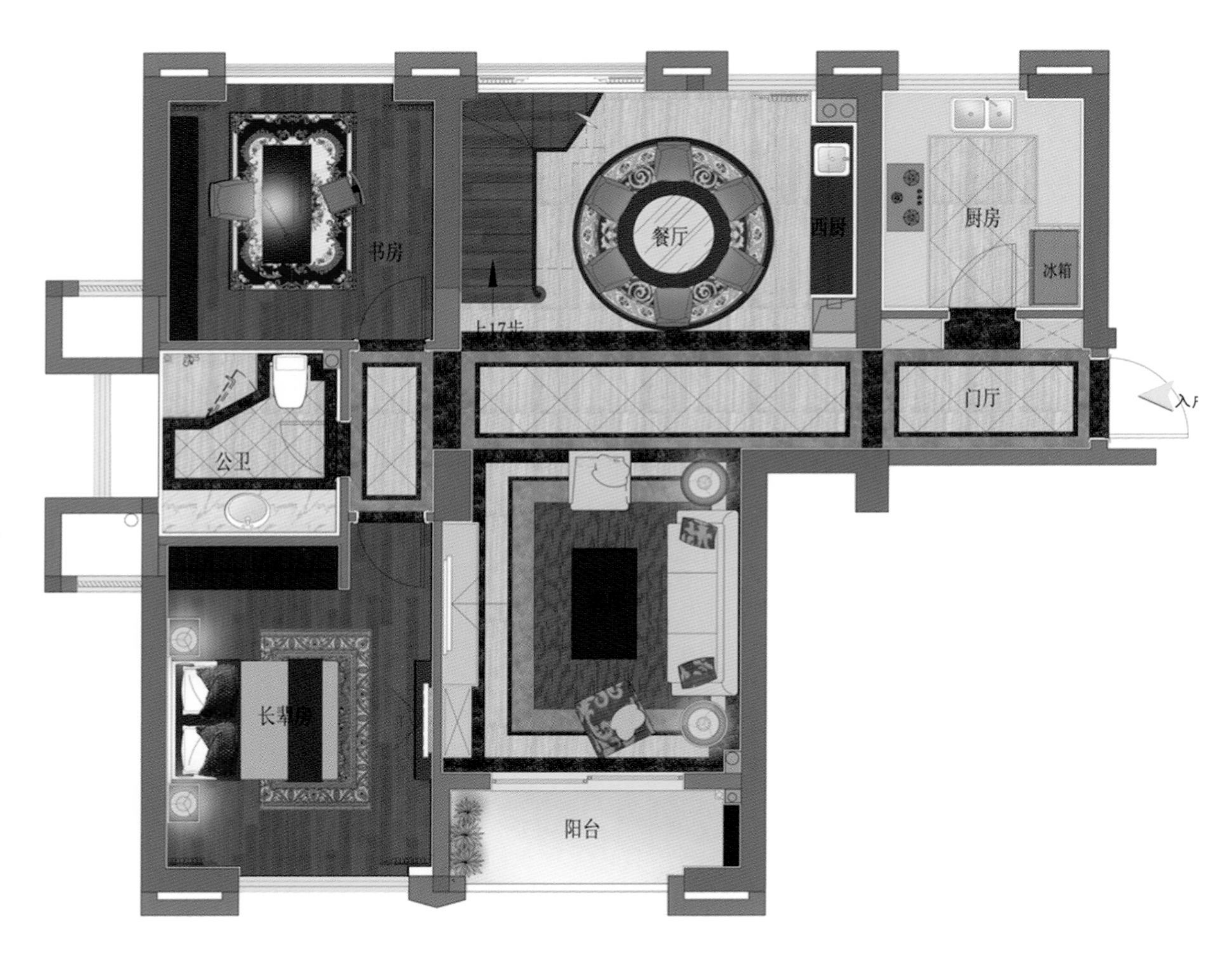

书房
上17步
餐厅
西厨
厨房
冰箱
公卫
门厅
入
长辈房
阳台

PGM
THE COUNT OF MONT CRISTO
THE COUNT OF MONT CRISTO

IT'S A GOOD DAY TO HAVE A GOOD DAY
A NEGATIVE MIND WILL NEVER GIVE YOU A POSITIVE LIFE
PEACE, LOVE AND JOY TO ALL WHO ENTER HERE.

17步
小坐垫
公卫
衣帽间
主卫
大女孩房
主卧
TV
阳台

本案雄踞浦上大道与乌龙江大道交汇处，坐拥乌龙江稀贵江景资源，高起点的地理坐标重新定义闽侯生活品质的新高度。平面布置上运用空间轴线的对位关系在主要的空间中让其有良好的展示性，在立面的处理上再将轴线所形成的对位关系打破，通过软装陈设的搭配让整个空间更加放松，自然空间色调以米白色、灰色、咖啡色为主，大面积降低色彩纯度，局部使用高饱和度色彩提亮空间。陈设色调使用 2016 年春夏色彩流行趋势——灰蓝、琥珀绿色、爱马仕橙。复古的美式和简约的现代风在空间中的碰撞，也别有一番趣味。

图书在版编目（ＣＩＰ）数据

空间美学 : 样板房风格设计透析 / 深圳视界文化传播有限公司编. -- 北京 : 中国林业出版社, 2017.1
ISBN 978-7-5038-8899-1

Ⅰ. ①空… Ⅱ. ①深… Ⅲ. ①住宅－室内装饰设计－作品集－中国－现代 Ⅳ. ①TU241

中国版本图书馆CIP数据核字(2016)第320579号

编委会成员名单
策划制作：深圳视界文化传播有限公司（www.dvip-sz.com）
总 策 划：万绍东
编　　辑：杨珍琼
装帧设计：潘如清
联系电话：0755-82834960

中国林业出版社 • 建筑分社
策　　划：纪　亮
责任编辑：纪　亮 王思源

出版：中国林业出版社
（100009 北京西城区德内大街刘海胡同 7 号）
http://lycb.forestry.gov.cn/
电话：（010）8314 3518
发行：中国林业出版社
印刷：深圳市雅仕达印务有限公司
版次：2017年1月第1版
印次：2017年1月第1次
开本：235mm×335mm，1/16
印张：20
字数：150千字
定价：398.00元(USD 79.00)